TRANSACTIONS

OF THE

AMERICAN PHILOSOPHICAL SOCIETY

HELD AT PHILADELPHIA
FOR PROMOTING USEFUL KNOWLEDGE

NEW SERIES—VOLUME 52, PART 4
1962

MEDIAEVAL ARABIC BOOKMAKING AND ITS RELATION TO EARLY CHEMISTRY AND PHARMACOLOGY

MARTIN LEVEY
Yale University

THE AMERICAN PHILOSOPHICAL SOCIETY
INDEPENDENCE SQUARE
PHILADELPHIA 6

SEPTEMBER, 1962

Reprinted 1966

Library of Congress Catalog
Card Number: 62-15117

PREFACE

Very little effort, in the past, has been made to study the relationship of Arabic chemistry and alchemy to the truly original and remarkable development of its contemporary, pharmacological science. To this end, the materials of a significant text on chemical technology have been studied together with those found in pharmacological texts, both in manuscript and in print. Where possible, names of botanicals and other materials of the technological text have been traced in the literature of chemistry and the materia medica from the earliest literate times down into the Arabic period and later. The etymology of these terms and their geographic distribution are of interest particularly for study of the paths of transmission of science from 3500 B.C. on.

Further, this study from source materials was made to assist the student of mediaeval science in his understanding of the scarcely known technological aspect of Arabic culture. The translation, as much as possible, is given with the view that it is the techniques and materials used which must, above all, be clarified. It is for the professional Arabist to carry on further detailed work in the philological difficulties of Arabic technological vocabulary exhibited in this text, as well as in the peculiarities of North African dialectal expressions of the eleventh century.

The major manuscript studied, like most of the others used, is frequently lacking in diacritical marks, thus making it difficult to ascertain the reading and meaning of rare technical terms. In transliterating Arabic words, the Library of Congress system has generally been followed. In the case of the diphthongs, *au* and *aw* have been used interchangeably. The same is true for *ai* and *ay*. For simplification, the plurals of *raṭl* and *mithqāl* have been anglicized but left in italics. The word dirham, since it is so common, has not been italicized.

Value of weights. The grain is a unit of weight. The carat equals 4 grains. A dirham equals 16 carats and is slightly less than 3.2 grams. Seven *mithqāls* equal 10 dirhams. An ounce equals 12 dirhams. A *raṭl* equals 12 ounces. One *dāniq* equals 1/6 of a dirham. One *istār* equals 4.5 *mithqāls*.

Where a reference book or article has been divided into sections and each section numbered, the reference given refers to this number. Otherwise, the reference is to the page.

It is a pleasure to acknowledge the many kindnesses and the expert assistance given to me by Professor Noury Al-Khaledy. Mlle Berthe van Regemorter was kind enough to give me helpful suggestions in regard to the art of bookbinding. To Professor G. K. Bosch, I owe thanks for allowing me to read her manuscript on the twelfth chapter of the ibn Badis text before its publication. Professor Bruno Kisch shared some of his expert knowledge of weights with me as well as his library. Dr. Jerry Stannard was very kind in reading the typescript. To Mary McGlinchy Levey belongs the credit for many ideas which are now unidentifiable because of their complete absorption and diffusion throughout the work. I am also grateful to my colleagues in other than mediaeval science who in many conversations and in correspondence offered assistance in many fields upon which this work touched and in other tangential matters. I am indebted to Temple University for assistance, and to the American Philosophical Society for grants for research which always seemed to come at an urgent moment in the course of the work. Completion of this work would have been very difficult without the support of the National Science Foundation and the National Institutes of Health (RG 7391). Finally, I wish to thank the Institute for Advanced Study for a year of quiet study where I thought about and commenced this work.

I dedicate this work to Susan Levey and to Peter Levey.

M. L.

CHEMICAL TECHNOLOGY IN MEDIAEVAL ARABIC BOOKMAKING

MARTIN LEVEY

CONTENTS

INTRODUCTION

1. TECHNOLOGICAL WORKS IN ISLAM

Study of the technical arts of mediaeval Islam has received scant attention compared with other branches of this rich culture.[1] One of the major reasons for this circumstance is that historians of technology are often insufficiently versed in the necessary range of languages and philology. Another cause lies in the difficulty that relatively few works have survived from the Golden Age of the Muslims, approximately ninth to twelfth centuries.

Much of the Arabic chemical literature is alchemistic in nature. Because of this, the student of the manuscripts is given a distorted view of the actual chemical knowledge of these people. There is, however, a small number of texts, in sharp contrast to the alchemical writings, which are devoted to the crafts of the time. These deal directly with or are tangential to technology and are of importance comparable to that of the alchemical works. Owing to their highly practical disposition, a much clearer view of the extent of Muslim chemical knowledge is thus imparted to the reader.

The latter type of literature is represented by the presently studied work of Muʿizz ibn Bādīs, *ca.* A.D. 1025, on bookbinding. Although the treatise is deficient in some aspects of the subject, it is, nevertheless, a rich repository of practical chemistry in the fields of tanning and dyeing of leather, manufacture of glues and their uses as adhesives and vehicles for paints, and the preparation of inks of many types for application to different writing materials.

Another work is given in translation in the appendix. It is by a master craftsman and is devoted largely to bookbinding. Since the ibn Bādīs manuscript has little on the actual art of bookbinding, the addition of this work should prove to be complementary to the former work. Its author is al-Sufyānī, a North African.[2] It was written in A.D. 1619. In addition to bookbinding, there are included descriptions of covering of the boards with leather, solution of gold and its application, manufacture of glue, dyeing of leather, and engraving of leather.

2. THE EARLY ISLAMIC BOOK

The codex form of book (*muṣḥaf*, from the Ethiopic) was known in early Arabic times. In fact, tradition has it that during Muhammad's lifetime, the Koran pages were kept between wooden boards (*lauḥāni* or *daf-fatāni*). Later Zaid ibn Thābit, before the end of the seventh century, copied it into book form.[3]

Fortunately, the Arabs in Ṣa'da, South Arabia had

[1] For a general discussion of the lack of cultivation of the study of the history of technology, *cf.* White, Lynn, Jr., Technology and invention in the Middle Ages, *Speculum* **15**: 141–159, 1940.

[2] *Cf.* review of P. Ricard's text in *Revue Africaine* **61**: 153–154, 1920.

[3] Arnold, T. W., and A. Grohmann, *The Islamic book*, 30, New York, 1929; Erpenius, Th., ed., *Al-Makīn, Ta'rīkh al-muslimīn*, 18, Lugd. Batav, 1625; Sarre, F., Islamic bookbindings, London, 1923; Adams, P., Turkish, Arabic, and Persian bookbinding, *Archiv für Buchbinderei*, 1904 ff.; Gratzl, E., *Islamische Bucheinbände*, Leipzig, 1924.

an excellent and flourishing leather industry.[4] Large tanning works, a necessary concomitant for a bookbinding industry, were also to be found in Kharran, Jurash, Ṣanᶜā, Zābid,[5] Bukhāra,[6] Egypt,[7] Maghrib,[8] and Spain.[9] It was in Ṣanᶜā where the white and yellow striped Cordovan leather was produced to rival the famous Moroccan leather of al-Ṭā'if. Ṭā'if leather is mentioned in one of the texts to be discussed.[10] Not only was leather a major product of this city but the region was known also for its excellent bindings.[11]

It was recorded in 985/6 that al-Muqaddasī, the bookbinder, received two dinars for a binding of the Koran in Yemen. In the Muslim Empire, bound books were very expensive although copying is supposed to have been done very quickly.[12] Bookbinders, booksellers, papermakers, and others who relied on the book trade flourished in this period. The bookstores were frequented by the intelligentsia just as in earlier and later times.

The Fihrist, in 987/8, mentions bookbinders who were well known.[13] The famous practitioners of the art listed are ibn abī al-Ḥarīsh, who bound books in the library of al-Ma'mūn (early ninth century), Shifat al-Miqrād al-ᶜUjaifī, Abū ᶜĪsā ibn Shīrān, Dimyāna al-Aᶜsar ibn al-Hajjām, Ibrāhīm and his son Muḥammad, and al-Ḥusain ibn al-Ṣaffār. The bindings of these men have been lost.[14]

Millions of volumes probably existed in the Golden Period of the Muslims. In Cordova alone, the library boasted of 250,000 books. Owing to the depredations of various groups many libraries were burned and otherwise destroyed. The library of the Fatimids in Cairo was sacked in A.D. 1068. It was not the first time this library was mutilated. It had been done a number of times previously by Christians and Arabs.[15] The library at Tripolis was burned by the Crusaders, that of the ᶜAbbāsids in Baghdad by the Mongols. The library of the Grand Master of the Assassins in Alamut burned in A.D. 1257 as did that in Medina in 1237.

3. SOME MANUSCRIPTS USED IN THIS WORK

The manuscript (1) upon which this study is mainly based is in the Oriental Institute, University of Chicago, catalogued as A12060. This is a copy made in 1908, by M. Sidiqi, a scribe in the Kutubkhane Khedive, of a manuscript then in the library of Ahmad Beg Taimūr and now in the Egyptian Library in Cairo. It was purchased by the Institute from Dr. B. Moritz in Berlin in 1929. It is equivalent to MS Landberg 637 in Berlin which is dated 1228/1813 and is called *Umdat al-kuttāb wa'uddat dhawī al-albāb*. Its author was al-Muᶜizz ibn Bādīs (*ca.* A.D. 1025). The full title is "Book of the Staff of the Scribes and Implements of the Discerning with a Description of the Line, the Pens, Soot Inks, Līq, Gall Inks, Dyeing, and Details of Bookbinding." Only the first ninety-three pages are relevant to the title. There are other copies of this manuscript,[16] in whole or in fragmentary condition. Those examined were (2) Gotha 1354 (67 fol.), (3) Gotha 1355 (56 fol.), (4) Gotha 1356 (1 fol.), and (5) Gotha 1357 (13 fol.). Some of the difficulties have been clarified by explanations in a variant manuscript, (6) A29809, of the Oriental Institute. This copy was completed in January, 1671.[17]

The author of the text, ibn Bādīs (A.D. 1007–1061),[18] a royal patron of the arts was born in al-Manṣūriyyah, near Qairawān, of the North African Dynasty of Zirides. He was a powerful and high-minded prince, a friend to the learned, and prodigal of gifts.

The other manuscript of which a translation is given is (7) *Ṣinaᶜat tasfīr al-kutub wa-ḥill al-dhahab* by abū al-ᶜAbbās Ahmed ibn Muḥammad al-Sufyānī. This text in 1919 was published in Fes (19 pp.) and a second edition in Paris in 1925 by Prosper Ricard.[19] The manuscript is dated 1029 A.H./A.D. 1619. The author, un-

[4] Müller, D. H., ed., *Al-Hamdānī, Ṣifa Jazīrat al-ᶜArab*, 113, Leiden, 1884–1891.

[5] Grohmann, A., *Allgemeine Einführung in die arabischen Papyri* 1, *Corpus Papyrorum Raineri*, Series Arabica I, part I, 51, Wien, 1924. In Ṣanᶜā alone there were at least thirty-three tanneries at the end of the tenth century A.D.

[6] Amedros, H. F., and D. S. Margoliouth, transl., ibn Miskawaīh's (d. 1030) *Kitāb tajārib al-umam*, in *Eclipse of the ᶜAbbāsid caliphate* 1: 223, Oxford, 1920/1.

[7] Becker, C. H., *Beiträge zur Geschichte Ägyptens unter dem Islam* 1: 182, Strassburg, 1902/3.

[8] Fumey, E., Al-Salāwī's Kitāb al-istiksa' li-akhbār duwal al-maghrīb al-askā', *Mission Scientifique Märocaine* 10: 59, Paris, 1906/7; Sprenger, A., Die Post und Reiserouten des Orients. *Abhandlungen f. d. Kunde d. Morgenländes* 3: 149 ff. 1864.

[9] Grohmann, A. Bibliotheken und Bibliophilen im Islamischen Orient, *Festschrift zum Zweihundertjahrigen Jubiläum des Bestandes des Gebaudes der National-bibliothek in Wien*, 431–442, Wien, 1926.

[10] Ibn Bādīs in chapter 12 gives the method of tanning of leather in Ṭā'if; de Goeje, M. J., *Bibliotheca geographorum arabicorum*, 1: 24, Leiden, 1870.

[11] Abu'l-Qasīm Ḥusain ibn M. al-Rāghib al-Iṣfahānī, *Muḥāḍarāt al-udabā' wa-muḥāwarāt ash-shuᶜarā'* 1: 70, Cairo, 1287. A fine binding of Kufic vellum in two boards done at Ṭā'if is mentioned.

[12] de Goeje, *op. cit.* 3: 100; *Fihrist* 1: 264.

[13] *Fihrist* 1: 10.

[14] For the famous calligrapher and bookbinder, Abu'l-Ḥasan ᶜAlī b. Hilāl of Baghdad (d. 413 A.H./A.D. 1022), see Rice, D. S., *The unique ibn al-Bawwāb manuscript in the Chester Beatty Library*, Dublin, 1955.

[15] Hammer-Purgstall, J. v., *Geschichte der Ilchane*, Darmstadt, 1842; *cf.* also his Übersicht der Literaturgeschichte der Araber, *Denkschr. d. K. Akad. d. Wissensch. in Wien, Phil.-Hist. Kl.* 2: 48, 1951.

[16] *GAL* SI 473,963.

[17] For physical description, *cf.* Levey, M. M. Krek, and H. Hadad, Some notes on chemical technology in an eleventh century Arabic work on bookbinding, *Isis* 47: 239–243, 1956. It has been mentioned by Bosch, G. H., *Islamic bookbindings; twelfth to seventeenth centuries*, Univ. of Chicago thesis, chap. 2, Chicago, 1952; and Karabacek, J., *Das arabische Papier*, Wien, 1887.

[18] deSlane, B. MacGuckin, ed., *Ibn Khallikan's biographical dictionary* 3: 386–388, Paris, 1868; same editor, *Ibn Khaldun, Histoire des Berbères et des dynasties Musulmanes* . . . 1: 20ff., Alger, 1847.

[19] Basset, René, *Revue Africaine* 61: 153, 1920.

fortunately, has not been able to procure a copy of the 1925 edition. The 1919 edition has been used in this study.

Further manuscripts have been examined on this subject—(8) Berlin 5565 (Sprenger, 1939). No title is given nor is the author listed. Siggel[20] believes that it is a fragment of a "manuscript of Zain ad-Dīn ᶜAbdur-raḥmān b. a. Bakr ad-Dimashqī al-Jawbarī," and is similar to MS Berlin 5563 (MS Wetzstein 1656) entitled *Kitāb al-mukhtār fī kashf al-asrār wa-hatk al-astār* and Gotha 1374.[21] The latter is called *Kitab al-mukhtar fī kashf al-asrār* and is by the same author.

(9) Berlin 5567 (Wetzstein 1375). The title of this work is *ᶜUyun al-haqā'iq wa-īḍāḥ al-ṭarā'iq* by abū al-Qāsim Aḥmad al-ᶜIraqī. Copy is dated 963 A.H./A.D. 1556.

(10) Yale L379. It is entitled *Al-nujūm al-shāriqāt* by M. ibn abī al-Khair al-Urmayūnī. It is of the sixteenth century but the copy is dated 1715. A variant of this manuscript (11) Landsberg 379 was also consulted.

The following fragments which include descriptions of the preparation of ink have also been utilized: (12) Gotha A 1327, (13) Gotha A 1358, (14) Gotha A 1349, and (15) Gotha A 1162.

(16) A manuscript which proved to be of value because of the large number of plant substances mentioned is that authored by al-Kindī called *Aqrābādhīn*, Aya Sofya 3603.[22]

(17) Another text similar in value to the preceding is the Cod. Or. 576 Leiden, *Minhāj al-bayān fīmā yastaᶜmiluhu al-insān* by ibn Jazla (d. A.D. 1100).

(18) *Al-dustūr al-bīmāristānī* by Abū al-Faḍl Dā'ūd b. a. al-Bayān al-Isrā'īlī (13th cent.) Münich 808, Gotha 2031, and text published by Sbath in *Bull. de L'Inst. d'Égypte* **15**: 13–78, 1933. Translation and commentary of this work are being prepared for publication by the author. References are to the page numbers of the printed text of Sbath.

(19) *Fī al-adwiya al-maujūda fī kull makān* by M. ibn Zakarīyā' al-Rāzī. Not foliated. Cushing collection, Yale University.

Other manuscripts used are given in the notes.

4. BLACK INKS

Ibn Bādīs divides his recipes for black inks into soot inks (chap. 2) and tannin inks (chap. 3). The former include the preparation of Chinese, India, Kufic, Persian, Iraqi, and Nafuran inks. The major difference among these inks is the material from which the soot is prepared. These are generally obtained from various botanicals. In the case of India ink, the important ingredient is a black sublimate from a mixture of vegetable and animal oils. Gum arabic is a common additive; glair is also used particularly in Iraqi ink. Very mildly acidic solutions of dilute vinegar or yoghurt were used to arrest or slow down formation of mold. This must have been a very serious problem to the Arabs in North Africa.

No mention is made in the text of the more refined Indian and Chinese method in which plants were burned with a limited access of air and the smoke conducted through long tubes of paper. The soot deposited at the end of the tube farthest from the fire was collected as the finest product for ink. In a few cases later on, notably with the heating of sulphur for its soot, recipes call for the pot on whose bottom the soot is to be deposited to be placed very close to the fire so that a minimum of air is circulated under it.

The most common black pigments are not mentioned. These are lampblack of various sorts and a natural black earth. Their preparation was too well known to be discussed.[23]

The gallnut and ferrous sulphate inks described are still used today. This type of blue-black was highly developed in a much earlier period for papyrus. Ibn Bādīs describes the preparation of many different types of tannin ink. Tannin is obtained mainly from the gallnut of terebinth and tamarisk. The vitriol with which it is used comes from such distant countries as Egypt, Cyprus, and Persia. Gum is generally used, sometimes with glair.

A number of specialized inks include an ink which can be used immediately after preparation, a dry ink and an ink for travelers, a cheap ink for the common people, an ink which does not require fire for its preparation,[24] and inks for religious books. In one recipe, pomegranate rind in addition to gallnut is used to prepare ink.

5. COLORED INKS

Black inks were almost always used by the Muslims for writing. Colored writing fluids for pen and brush were employed for rubrics, capitals, flourishes, lines, line finishing, embellishments on page borders, and pictorial decorations. Some of the colored inks were used on leather. Although some fine miniatures are to be found in Arabic manuscripts, there were not many since the pictorical art was not only discouraged but repressed to a large extent among Muslims. There can be no doubt that many of the colored inks described in this chapter served mainly in the early illumination of the beautiful Arabic script and decorative designs in

[20] Siggel, Alfred, *Katalog der arabischen alchemistischen Handschriften Deutschlands: Hss. der Öffentlichen Wissenschaftlichen Bibliothek, Früher Staatsbibliothek Berlin*, 125, Berlin, 1949.

[21] Siggel, A., *Kat. d. ar. alch. Hss. Deutsch: Hss. der ehemals herzog. Bibliothek zu Gotha*, 86, Berlin, 1950.

[22] I am indebted to Professor Fuat Sezgin of Istanbul University for calling my attention to this important work. It is being prepared for publication.

[23] Thompson, D. V., Jr. and G. H. Hamilton, eds., *De arte illuminandi* (an anonymous fourteenth-century treatise), 2, New Haven, 1933.

[24] Wood and other fuels have been scarce and expensive in some parts of the Near East for many years.

the Muslim book. It may be inferred that the brush primarily was meant to be used with the paints in this chapter. They could not have been used with the *līq* preparations in the next chapter.

Colored inks, according to ibn Bādīs, fall into three groups, red, yellow, and green. Although other colored inks are described such as violet, white, blue, and pink, these three were evidently considered the most important. Peacock blue ink made with gum is mentioned. Blues which are transparent tend to be lean and opaque when laid with glair. There is nothing in the manuscript to indicate that the author was aware of this. In Latin texts of the eleventh century and later, the well-established practice was to temper blues with glair. Gum tempering was less generally recommended than tempering with glair.

In later mediaeval times, gum tended to displace glair. The gum was largely used for brush work while the glair was for tempering pen colors. Not many glair recipes are found in ibn Bādīs. The use of some type of gum was much more common. The compromise of the use of glair plus gum is more apt to be found in ibn Bādīs than glair alone.

Some of the other pigments used are of interest. One of these, "yellow arsenic," is orpiment. Although orpiment, As_2S_3, is found in nature, its preparation was well known to the alchemists. Cennino in his work, *Il Libro dell'Arte o Trattato della Pittura* (early fifteenth century),[25] states that orpiment is artificial, meaning that it was made by the chemists of his day.[26] Other golden colors are described by ibn Bādīs.

Verdigris was a very common green pigment, in fact, it was one of the basic materials of illumination in the early mediaeval period. In ibn Bādīs, this is the favorite green in spite of the fact that botanical greens are much better since verdigris is much more reactive with the commonly used orpiment.[27] Holmyard[28] has shown that Mary the Copt in Egypt (*ca.* third or fourth century) knew that vinegar and copper produce the pigment verdigris. An exact procedure for its manufacture which would control the variation in the color of verdigris is not provided by the ancient literature.

The two most important red pigments were cinnabar and red lead. Classical writers as Pliny and Isadore used *minium* for cinnabar as well as for other red materials. (*Minium* is the source of the Latin and Italian *miniare*, "to miniate," "to rubricate" manuscripts. *Miniatura* then gives the English "miniature.") Much confusion is found in the Latin terminology for cinnabar. The preparation of red lead, Pb_3O_4, is well known in many manuscripts. Its orange color when pure is very difficult to preserve.[29]

It is of further interest that some recipes contained lac. Gum, however, is always present. It seems to have been used partly for its rich lustrous finish. The purpose of a solution of gum or oil in ink is both to hold the writing in a state of suspension and to act as an adhesive. This was not always clearly understood so that gum is found in recipes for inks which are active chemically and so do not require an adhesive. In the case of papyrus, parchment, and certain sized papers, the ink frequently acted as a paint. In this event, the gum was essential. Less important properties of gum are that it retains a certain viscosity for a time when proper care is taken and that it serves to protect plant derived matter from more rapid decomposition.

Some of the ink recipes in this chapter contain gallnut in spite of the fact that the tannins have a deleterious effect on the colors although a small amount prevents the formation of mold.

6. *LĪQS*

The *līqs*, ink-soaked wool or felt wads, were meant for use with the pen. They were made in a greater variety of colors than those pigments prepared for the brush. The inks for *līq* were prepared in essentially the same manner as were those used for the brush.

Gold-colored inks were very common. These were prepared from the yellow pigments of various botanicals.

There does not seem to be much reason for the ingredients in some recipes. For example, in the preparation of a white *līq*, to white lead and mica are added gum arabic, gum tragacanth, and fish glue. It would not only be difficult to use but almost impossible to keep. Another recipe has the two gums together with lac.

It may be recalled that gum tragacanth swells enormously in water so that when it is used as a tempera the colors should be bound with a minimum of it. The proportions in the recipe do not bear this out. Gum tragacanth contributes little to the optical effect but is effective as a binder. It is likely that ibn Bādīs himself was not a practicing craftsman in this art.

7. MIXTURES OF DYES AND COLORS

For mixtures the basic colors are given as white from white lead, black from soot ink, red from cinnabar and red lead, green from verdigris, yellow from orpiment, and red from realgar.

The text is concerned not only with the mixing of these so-called basic pigments but also with their tint-

[25] Herringham, C. J., *The book of the art of Cennino Cennini*, chap. 47, London, 1922; original text edited by D. V. Thompson, Jr., *Il libro dell'arte*, New Haven, 1932.

[26] The Leiden and Stockholm papyri are concerned with dyes. Many of these were intended to imitate the colors of gold and silver. Berthelot, M., *Intro. à l'étude de la chimie des anciens et du moyen-âge*, 1–73, Paris, 1889; Berthelot, M., *Les origines de l'alchimie*, 80–94, Paris, 1885. These papyri are *ca.* A.D. 300.

[27] Thompson, D. V., Jr., and G. H. Hamilton, eds., *De arte illuminandi*, note 26, New Haven, 1933.

[28] Holmyard, E. J., The letter of the crown and the nature of creation, *Archeion* **8**: 161–167, 1927; Thompson, D. V., Jr., Artificial vermilion in the middle ages, *Technical Studies* **2**: 64–65, 1933.

[29] *Cf.* note 41, Thompson and Hamilton, *op. cit.*

ing by the use of white lead "to give a multiplicity of dyes," or by adding yellow arsenic or indigo little by little.

Mention of bone-white is not made at all. In the Middle Ages, it was sometimes used in place of white lead since the latter was so reactive with orpiment and also verdigris. The anonymous author of the *De arte illuminandi* stated, "... but it is not a good plan to use orpiment on parchment, because by its odor it reduces white lead, red lead, and green to a sort of metallic color...."[30] Bone-white, however, is a poor pigment in covering power and is bulky to use.

A medium is not always used in the tinctures described. This may possibly be due to the fact that gums would tend to mask a slight change in coloration.

8. METALLIC AND SECRET INKS

For calligraphic writing, the color and luster of metals is frequently desired. Metallic inks are prepared in two ways, (1) by the use of actual metals and (2) coloring matters to which a metallic luster is imparted by special treatment.

Gold ink was made by ibn Bādīs by using finely pulverized gold, from leaf or filings, in various media which do not detract from the metallic luster. Gum tragacanth is an excellent vehicle in one recipe.

Silver ink is made with silver in essentially the same way. A substitute for silver in a recipe is given as tin (or lead). The amalgam is formed and is then pulverized; the gums are then added. Another recipe for a silver substitute includes unslaked lime and glue. The best metallic inks in the nineteenth century were still made with pulverized gold or silver in the same way.[31]

In the mediaeval period there were countless rules for gilding.[32] Gold size is not mentioned by ibn Bādīs nor did he know of work with a mordant whose purpose it is to lay a determinate shape in unburnished gold.[33]

The text contains a number of recipes which are unrelated to metallic pigments and belong to an earlier section.

There are two recipes for copper inks. Copper inks, of course, cannot retain their color for long in the atmosphere.

Sympathetic inks are generally defined as those fluids which when subjected to a certain treatment, either change color, vanish, or appear. The recipes given by ibn Bādīs belong to the last category. They include the use of yoghurt, sal ammoniac, milk, and white vitriol as the primary inks. The secondary materials are gallnut solution, ashes of paper, and boxthorn. Heat is sometimes required. These inks are of little practical value.

[30] Thompson & Hamilton, *op. cit.*, 7.

[31] Lehner, S., *The manufacture of ink*, 145, Phila., 1892; Davids, T., *The history of ink*, New York, 1960.

[32] Chapter 14 of *De arte illuminandi.*

[33] *Ibid.*, chap. 31.

9. ERASURES FROM PAPER AND PARCHMENT

Many of the recipes in this section of the text are for merely covering the spot desired in the manuscript. Vinegar solutions, yoghurt with salt in wool, and salts are used to remove inks. A rough eraser is made of "*iqlīmiyā* scum from a melting metal" with the sharpness of citron. This is then employed to rub out the writing.

Most of the ink used by the Arabs was soot ink which did not react chemically but penetrated the paper physically. This ink is difficult to remove as is the tannin type. Since carbon ink was so difficult to discharge, the Arabs simply covered it, as they did with other inks with a type of paint preparation.

A manuscript[34] gives a recipe for erasure fluid.

> Wax is melted and then saturated with incense. This is used when necessary. Then, when you put it on the writing, smear it. As often as the spot is whitened, return to another spot until a trace of writing does not remain.

This was simply a covering procedure, not actual eradication. Another one different in effect is,

> Take Yemenite alum, *qalī* and sulphur in equal parts. Soak them in vinegar until dry. When dry, it is like dough. Work it on in layers. Leave it until it is wiped. Ammonia had been used earlier. Then the layers on the ink are removed. There does not remain a trace of the ink.[35]

10. GLUES, ADHERENCE OF GOLD AND SILVER, POLISHING

The two main types of glue used are fish glue and snail glue. In mediaeval times, it was not enough to use glue on the gold leaf itself. The gold size from fish and snail glues has sufficient gelatine for adherence to parchment. However, it may not be sufficient to insure against any chipping off under the burnishing. Therefore, gilding experts applied adhesive to the parchment or paper.

In some recipes throughout the text, honey is employed. A little of it is sufficient to keep the binders, as gum, size, and glair, slightly moist in order to avoid excessive brittleness when dry and to maintain a slight flexibility. This helps to prevent cracking. Sugar is a hygroscopic substance which can be employed for this purpose. However, in this text, it is fairly certain that sugar is used only for the formation of its tiny crystals which when dry give a sparkle to the ink.

Ibn Bādīs does not give definite instructions on the making of fish glue. This is because it was so common.[36] It was made from the skins, bones, and entrails of various species of fish.[37] Glue was also obtained from scraps of hides. The detailed procedure is given in the

[34] MS Berlin 5565, fol. 13*b*.

[35] *Ibid.*

[36] *Cf.* Cennino, *op. cit.*, chap. 108.

[37] Theophilus quoted in note 30, Thompson and Hamilton, *op. cit.*

text. An elaborate description of the preparation of glue is given by a craftsman, al-Sufyānī. See appendix.

For polishing Cennino[38] used haematite burnishers. The teeth, he says, of dogs, lions, wolves, cats, leopards, and of all clean carnivores are also good. These were well polished to round off any sharp edges.

11. MANUFACTURE OF PAPER AND ITS TINTING

Paper is manufactured in the text from flax. There is a fairly full description particularly showing the care exerted in the operations. It is one of the earliest descriptions in Arabic.[39] Paper, however, was known much earlier. Originally, paper came from China over a caravan route through Central Asia and Persia. At Samarkand, the route divided, one branch going to Kashgar and the other to Serinde. At the latter, paper from the sixth century has been discovered. Papermaking was later developed fully in the Islamic world when, in 751, Kao Hsien-chih was defeated and several Chinese papermakers were captured and put to work. Samarkand had abundant crops of flax and hemp as well as sufficient water.[40] From here paper came to the West although the transmission took about 500 years.[41]

What may be the oldest sample of European paper, although there is as yet incomplete proof, is a document dating from 1109 in Palermo. It is an order in Greek and Arabic concerning a salt mine near Castro Giovanni issued by the wife of Roger I of Sicily.[42]

According to ibn Bādīs, the flax is soaked in quicklime, rubbed with the hands, and spread out in the sun to dry. It is then returned to fresh quicklime. This is repeated a number of times. Then it is washed free of the quicklime many times, pounded in a mortar, washed, and introduced into molds of the proper measure. Care is exerted so that the thickness of the paper is regular. It is then left to dry. It is treated with rice water or bran water. Starch is also used for this purpose. It also helps to glaze the surface of the paper.

12. BOOKBINDING AND ITS TOOLS; COLORING OF PAPER; GLUE FROM LEATHER SCRAPS

Part of the text is concerned with the binding of books and the necessary tools.[43] There is a brief description. The leather for binding is declared to be more desirable when well tanned and soft. Procedures are given to correct improperly tanned leather.

When the leather is dyed, alum is used as a mordant. Tannin and iron compound are used to color leather black. Carthamus is employed together with *qalī*, a mixture mainly of sodium and potassium carbonates.

The coloring of paper was practiced in the time of ibn Bādīs. The text outlines a simple procedure. In a later manuscript,[44] it is given in more detail:

On the dyeing of paper. If you wish to dye paper red, then take ten dirhams of *lukk* and the same amount of soda. These are kept warm. Sweet water is poured on the soda, then cooked until half of it has evaporated. It is clarified, then the *lukk* is added to it. A small amount of *bauraq* is added, then cooked until half of it has evaporated. It is removed from the fire and cotton is soaked in it. The paper is coated with it, then polished. If indigo is mixed with red arsenic, paper soaked with it, and kept warm until dry, then it tends to a yellow color.

Another description. Cook *ḥitmīt* until it is green, then soak the paper in it until it is green. Polish its surface, keeping it warm in the atmosphere.

Another description for yellow. Macerate root of the *qīsāʿ*[46] until it becomes yellow. Then finish with a little urine. Soak the paper in it; polish it.

Paper is placed in a solution of mauve,[46] vinegar, and arsenic. Then it is polished. It comes out nicely. Or take ten dirhams of verdigris and a *dāniq* of saffron. It is mixed with water and the paper is dyed with it. It comes out the color of mauve.

For verdigris dyeing, take the best verdigris. It is put with bloodstone in a mortar, then washed and purified. What was purified is placed in the water to use as a dye.

For a rose colored dye, *lukk* is mixed with a little white lead, then used to soak the water in.

It is surprising that the coloring matter was not added to the pulp before it was put into the molds. Only the already molded sheet seems to have been dyed by the Arabs.

An interesting process is described for the manufacture of glue from leather scraps. Following an ancient Sumerian procedure, it involves depilation of the leather, soaking in water in a vat until disintegration, then heating and filtering. When cold, it is cut into small pieces.

Immediately following the end of the twelfth chapter where acknowledgment is given and prayers are made to Allah, a number of notes are given. These are concerned with the testing for genuineness of such substances as white lead, verdigris, mercury, opium, musk, and others employed in the text. Most of these tests to prevent falsification of the materials are very simple, easy to perform, and empirical in nature.

13. CHEMICAL APPARATUS AND PROCESSES

The chemical apparatus and processes in the ibn Bādīs text display a very definite similarity and link with that in the Arabic and Alexandrian Greek alchemy

---

[38] Herringham, *op. cit.*, chap. 135, 136.

[39] Karabaček, J., Neue Quellen zur Papiergeschichte, *Mitt. aus der Sammlung der Papyrus Erzherzog Rainer*, **6**: 79, 82–83, 1888.

[40] Blum, A., *On the origin of paper*, 19 ff., New York, 1936.

[41] The *Mishnah* (canonized *ca.* A.D. 300) mentions *neyar* which may be paper. *Mish. Sabb.* 8, 2 (78*a*).

[42] LaMantia, G., *Il primo documento in carta* (*Contessa Adelaide*) *esistente in Sicilia*, Palmero, 1908.

[43] For tools used in decoration, *cf.* Fache, J., *La dorure et la décoration de reliures*, 78 ff. Paris, 1958.

[44] Berlin 5565 (Sprenger 1939). ʿAbdurraḥmān b. a. Bahr ad-Dimashqī al-Jaubarī (1st half 13th century), *Kitāb al-mukhtār fī kashf al-asrār wa-hatk al-astār*.

[45] Probably *Rhamnus alaternus* L., Maim. 93, I.B. 1403.

[46] Probably *Corchorus olitorius* L.

as well as with that found in Babylonian chemical technology. Some of the apparatus employed by ibn Bādīs include:

Vessels

jarrah—a clay container
qillah—clay vessel that can be heated in a furnace
qawārīr diqāq—storage vessel buried in dung
qidr—a pot which can be heated
qinīnīah—flask
qamqam—a flask with a narrow width that can be heated
qarūrah zajāj—long necked flask
qarʿah wa-anbīq—cucurbit and ambix
ijjānah—basin, tub, amphora
ṭast or ṭasht—flat dish
barniyyah khaḍrā'—glazed pot
ḥuqqah—container used to dissolve glue
ṭaijan maṭliyy—glazed pan
faqqāʿah zajāj—glass used with a very wide mouth

Ovens and furnaces

furn—a baking oven
atūn—a furnace
atūn al-zajāj—furnace for glassmaking

Other apparatus

manākhil al-julūd—leather sieve
manākhil qārūṭ—miller's sieve
rāwūq—filter
gharbāl—a sieve
mighrafah ḥadīd—iron deflagrating spoon
qālib—mold
ṣalābah—stone on which pulverization takes place

With necessary exceptions, the apparatus in use by the chemical technologist in this text is reminiscent of that used by al-Rāzī in *Secret of Secrets*, an alchemical book,[47] and by al-Kindī[47a] in his book on perfumery. Thus the processes in use by the chemical technologist are similar to those in use by the alchemist. This is supporting evidence that the chemical technologist or master craftsman was aware of much of the knowledge of the alchemist. The insinuation here is that, at the beginning of the eleventh century, there was no gulf between theory and practice in chemistry.

It has already been demonstrated[48] that early Muslim chemists owed a considerable debt to Babylonian chemistry for its apparatus and processes. Islamic chemical technology is also under the same obligation.

14. CHEMICALS IN ARABIC TECHNOLOGY AND ALCHEMY AND THEIR SOURCES

The ibn Bādīs text demonstrates not only the relationship of two important lines of evidence, i.e. apparatus and procedures, but it casts a new light on another facet. The long list of chemical materials herein contained is a rich one for investigators of Islamic chemical technology not only for their occurrence in this unusual type of text but also because their uses are given in clear context. Most of the chemical principles employed were products of botanicals previously known only in the medical literature.

These have been studied in three directions. First, comparison of the pharmacological and other uses of these botanicals has been carried out in Greek, Indian, Babylonian, Arabic, and contemporary Near Eastern sources. Second, an abbreviated etymological study of the botanical terms was undertaken with consideration of the more important older Indo-European and Semitic languages, Turkish, Sumerian, Middle Egyptian, and Berber. The third direction of study was to determine, if possible, the geographic areas where many of these plants are known to have been indigenous. Finally, correlation of these results has been made to determine, if possible, the origin, the improved and more certain identification of the materials, and their paths of transmission to Islam.

The evidence given in the notes can support only a partial answer to these questions owing to some of the obvious and some not so evident pitfalls in an attempt such as this. However, a preliminary result tended to demonstrate that the early literature in Sumerian and Akkadian already possessed a large portion of the later Muslim technical terminology. Much of the chemistry of metals and salts had been well known to the Babylonians at an early date so that many of the technical terms, especially those in the Semitic Akkadian, have cognates or relevant translations in the later Arabic. There is thus a relative conservatism in the names for chemical and botanical substances and they frequently occur in some kind of morphemic and semantic patterning.[49] For example, the words for filter and distillation, as well as many of the metals and botanicals, fall into this group.

Other terms, to a lesser extent, come from or were from the same origin as Sanskrit, Persian, Greek, and a few other languages. Some of the terms were found to be similar to those in the geographical regions where the botanicals were indigenous showing outright borrowing. These influences and not the powerful effect of Greek writers such as Dioscorides, Theophrastus, and Galen upon Arabic science were found to be more predominant linguistically.

Many of the Arabic terms for botanicals were found to have uncertain meanings even though their uses, particularly in pharmacology, coincided in various chronological periods, before and during the Arabic period. Odoriferous materials in ancient and mediaeval

[47] al-Rāzī, 54–63.

[47a] al-Kindī, 18–21; Levey, M., A group of Akkadian texts on perfumery, *Chymia* **6**: 11–19, 1960. The latter shows the dependence of Arabic perfumery on that of the Babylonians.

[48] Levey, M., Early Muslim chemistry: its debt to ancient Babylonia, *Chymia* **6**: 20–24, 1960.

[49] For the scientific use of these relationships, *cf.* de Almeida, A., Contributions à l'étude de l'anthroponymie des Dembos, *Fifth International Congress of Toponymy and Anthroponymy: Proceedings and Transactions* **2**: 349–352, Salamanca, 1958.

times from India to Spain were thought to be valuable in many illnesses. Astringent substances were universally employed for "drawing flesh together" in wounds, boils, and other suppurative lesions. For example, myrobalan was prescribed for pustules of the mouth.[50]

Some of the simples were restricted to external use because they were poisons. Others, such as narcotics, were for obvious uses. Aside from remedies of the type mentioned, there was some measure of agreement in the pharmacological uses of a few drugs. Those which were employed for diverse types of ailments were more difficult to identify in many cases since their properties were not sufficiently delineated.

A statistical evaluation of the etymology of the thousands of chemical terms used by the Muslims will help shed a proper light on the three questions posed above. The previous discussion is meant primarily to bring to the attention of serious students the possibilities of linguistic and statistical applications to the history of ancient science.

Because of the need for an ancient materia medica, botanical studies were kept alive and flourishing; this was particularly true in mediaeval Islam. Owing to the scarcity of technological literature and the relative abundance mainly of medical writings among ancient and mediaeval peoples, it is possible that the one strong thread which helped retain much of the ancient knowledge of the materials used in chemical technology was this medical and allied literature. This literature together with that on the later alchemy, all worked toward the retention and strengthening of chemical technology. This, together with the father-son apprenticeship system, served to carry on the transmission of chemical technology.

15. TESTING THE PURITY OF MATERIALS

The last part of the ibn Bādīs text is concerned with a number of notes on how to determine the genuineness or falsity of the materials to be used in the work. Not only are these probed by their physical properties but an attempt at chemical testing is made. In early Arabic literature, the physical tests of color, odor, taste, and others, were commonly used. Ibn Māsawaih (A.D. 777–857), for example, in his *Treatise on Simple Aromatic Substances* employed physical but no chemical tests although he was much concerned with the purity of aromatics. Throughout the Arabic period, there was a development of chemical testing methods to ensure greater purity of materials. To a major extent, the motivating force was a commercial one.

Most of the chemical changes are carried out by ibn Bādīs with the aid of heat. Otherwise, they are performed by adding water or drying in the air. The physical changes of the resulting reactions a re observed

[50] For Sumer, *cf.* Levey, M., Sumerian medicine of the third millennium B.C., *Actes VIII Congresso Internazionale di Storia delle Scienze*, 843–855, Firenze, 1958; for the early Arabic period, *cf. Aqrābādhīn*, fol. 107*a*.

STAFF OF THE SCRIBES AND IMPLEMENTS OF THE DISCERNING WITH A DESCRIPTION OF THE LINE, THE PENS, SOOT INKS, *LĪQ*, GALL INKS, DYEING, AND DETAILS OF BOOKBINDING[51]

by

Al-Muᶜizz ibn Bādīs

In the name of Allah, the Merciful and Compassionate One.

First Chapter. The excellence of the pen, the line, the choice of the best pens, their selection and differences according to the appearance of their lines, description of the inkwell, and the choosing of its instruments from the knives and others.

Second Chapter. On the making of the kinds of soot ink.[52]

Third Chapter. On the preparation of the types of black tannin inks.[53]

Fourth Chapter. On the preparation of the types of colored tannin inks.[54]

Fifth Chapter. On the preparation of *līq*.[55]

Sixth Chapter. On the tinting and mixture of dyes.

Seventh Chapter. On writing with gold and silver and their substitutes.

Eighth Chapter. On putting down secrets in the book.

Ninth Chapter. On the preparation of erasure materials for the writing on paper and parchment.

Tenth Chapter. On the preparation of glue from snails, solution of fish glue,[56] adherence of gold and silver, description of its polishers and the polishing, hair pens, quill pens, and all the instruments for gold and silver.

Eleventh Chapter. On the preparation of paper,[57] decorating of pens and their engraving, soaking of paper, and its beautifying.

Twelfth Chapter. On description of bookbinding[58] in leather, the leather covers, and all of the instruments so that it will suffice for one to do his own bookbinding with leather.

FIRST CHAPTER
ON THE IMPORTANCE OF THE PEN AND WRITING

The blessed and exalted Allah referred to "the inkstand, the pen, and what they write." The Exalted One said, "Read, by your generous Lord who taught by pen." And the messenger of Allah, may Allah bless him and peace be upon him, says that the first significant thing that Allah created was the pen. **(2)*** And he said, "Flow," and it flows with whatever it is, until the day of resurrection. And ibn ᶜAbbās[59] said, in the word of the exalted Allah, "Entrust me with the treasures of the earth since I am a guardian and a learned one." This means a curator of records. And of the honor of the pen,[60] without which no book was written, ibn ᶜAbbās, with whom Allah, the exalted, was pleased, commented upon the Exalted One's word, saying, "the pretty writing." This is in a commentary on the Exalted One's work, "When they find their pens made of sticks, their names written on their tops [i.e. of the pens]." Some of the commentators have said that the Exalted One said that he adds in the creation whatever he wishes, thus beautiful writing. And the prophet, may Allah bless him and give him his peace, said, "Beautiful writing gives to truth more clarity. It demonstrates that when the pens are good, the books smile." The pen is the molder of speech. It molds whatever the mind contains. Anything which the pens have given fruit the ages have not dared to erase. But the pen is a tree and its fruits are the words and the thought is the pearl of wisdom.[61]

Description of the choice of the best pens,[62] their selection and the variation of their appearance in the

---

[51] A gloss reads: Staff of the Scribes and Preparation of the Discerning on Writing and Preparation of Soot inks, Dyeing and Gilding, and Solution of Gold.

[52] *Midād*, from *madd*, to stretch out (i.e. the ink). (This is not the same as that in medical literature; for example, in Diosc. (v: 162) there is a compound medicine made of root of pine wood, gum arabic, and oil.)

[53] *Ḥibr*, from *ḥabar*, to write. *Midād* indicates a paint type of ink whose important ingredient is soot. *Ḥibr* refers to an ink which reacts chemically with paper or parchment. It is usually a gall-vitriol type of ink. This distinction was often later blurred. ᶜAbd al-Bāsiṭ b. Mūsā al ᶜAlmāwī (b. 1502 in Damascus, d. 1573 in Saᶜbān) in his *Al-muᶜid fī ādāb al-mufīd wal-mustafed* an abridgement from M. al M. al-Badr al-Ghazzī (MS Damascus 1349) claimed that *ḥibr* was better than *midād*; Rosenthal, F., *The technique and approach of Muslim scholarship*, 9, 13, Roma, 1947; Brockelmann, C., *Gesch. d. arab. Lit. S.* **2:** 488, 1942.

[54] *aḥbār al-mulawannah.*

[55] *Līq* is a piece of wool or felt which is soaked in ink and placed in the inkwell. The pen is dipped into this. It is still used in the Near East. Its advantages are that it cannot be spilled by travelers or children, it cleanses the pen as it is dipped, and the ink is held in a "state of suspension."

[56] *ghirā.*

[57] *kāghid.*

[58] *tajlīd.*

* Bold-faced numerals in parentheses indicate page number of original document.

[59] Ibn ᶜAbbās was the son of Muhammad's uncle.

[60] *qalam* (*aqlām*).

[61] For the origin of this and other quotations emphasizing the importance of the art of writing as well as some important scribal authorities and artists, *cf. Fihrist*, 10.

[62] For preparation of the pen in Talmudic times, *cf.* Krauss, 155.

kinds of writing. Description of the inkwell, selection in its thinness, thickness, the inside length and shortness, and measure of its sides. One puts a split in it, that is the tip of the pen, below its middle and its head, as much as a thumb. Its splits are interrelated in thinness and pointedness. This split is from the middle two-thirds up to the head of the pen. The lighter in weight, the more artful the pen. (3) If it is short, then it is thicker and stronger. The preferred length should have oil in it. No corner is left; so nothing may be gathered on the thick nib[63] on the inside and the edges. If the pen is even, the writing comes out thin and not so pretty. If it is diagonal, the line comes out weak. The most beautiful and best of the properties is the goodness of the average one—between the long and short, the thin and the thick, the diagonal and straight, the edge and the inside. It has more resemblance to the writing by ink on paper and copy books. As to any other than these, they are not so desirable. The best of the reeds is that which is proportioned in its length, its body, and its hardness. The chosen one is that which has a redness within it and is more oily. It is necessary for the preparation of this pen that it be sharpened from its head, that is, the thick part of the reed. If the contrary, then it is weak. So, it is necessary that the pen be sharpened from its lower point since it is stronger than its head. This is the fine section of the reed. Be aware that it is not so adaptable for the diagonal one as it is for the straight one when it is in the hand of the writer. Therefore, it is essential that the nib of the pen be straight.[64] It has a corner by the right split section. It may be thought of as having edges. It is necessary that there be a split of the pen in the middle. The distance of the nib to approximately the place where the pen is held, is as long as the distance of the joint of the small finger to its head (i.e. the head of the finger[65] to its first joint). What one usually sharpens and writes with is a quill. For the thickest pen, some of the fat in its head is removed. Only a little remains. If the fat from the beginning to the end is of the same consistency, then the flow of the pen is easy. Otherwise the line is not beautiful and it is bad. If its head has much fat, it does not write. (4) It is necessary that this be taken into account.

The writing of the straight pen is stronger and smaller and is more permanent. According to the opinion of the scribes, it is better and more beautiful. The writing of the diagonal pen is weaker than others but more beautiful; it resembles writing on paper. The middle one between them has both characteristics. The length of its head assists the light hand in speed of writing. That which is shorter is just the opposite of that. If the sharp edge of the pen is long, its writing is lighter and weaker. If it is short, its writing is stronger and heavier. What is selected and recommended is the medium one in the three cases. This is the one which is average, between the long and short, its thinness, diagonality, and roundness. The best of the pens is that prepared by shaving away its sides in the middle so that the pen point is a little wider than the middle. The length of its edge is as much as the pen. More than that or less spoils it. The man puts the knife on the reed straight. His hand should be not to the right nor to the left, not crooked or reversed so that it does not go a little to one side. It is done with the right hand grasping the knife so that it can cut. It causes the knife to cut on an angle and not perpendicularly. Otherwise, the pen is nicked and the parts are jagged. Putting it in the middle safeguards the two edges of the pen. Scrape[66] it off slowly, little by little, as a toothpick is scraped. The oil of the pen is average, neither thick nor thin, so that the going is easier for the pen. (5) If its oil is in excess, the pen is slow. If it is thin, it flows weakly.

If you begin to cut the pen, then cut it vis-à-vis the plant of the reed, i.e. the small hole in the lower part of the reed. It seldom is spoiled if done that way. If you wish, scrape off the pen. Do not work on the two sides at once, not with the middle, and not with the fat. Take the knife and scrape off one side. It will take a long time to make it straight; you will have to stop. Then, begin first with the middle of the two edges so that twisting is avoided and its form remains true. Its lower part becomes shortened and then it is the right that is fuller than the left. This is a necessity for writing. If the left is fuller than the right side, then it will spray and so ruin the writing. While you split the pen, keep firm. Do not make haste or you may slip from the proper path. The value of the pen depends upon the correctness of its split. This has been described. The same is true for its cutting. The true way is for the right edge to be full while the left edge is a little less so. If you prepare it according to what has been described, then cut it medially, neither long nor short, tending, however, toward the long. This is the choice of all writers. If this is so, then it is correct in its flow. When you cut it, cut it straight. It is necessary that the pen be cut while the edges are together. It is then opened. The writing then is more beautiful. (6) If it is cut and it opens a little, then it is that it may split. This spoils it. If its opening becomes very wide and it is cut after that, then the damage has been done. For this reason, the pens of the common people are damaged. They do not know how to cut the pens for they have no idea of it. Perhaps they may cut it after they write with it. That happens to those who do not care for beauty and straightness of writing and the perfection of this art.

Description of the knife for pen cutting.[67] As to the knife it is essential that it be of iron, the best and the

[63] *qaṭṭ*.

[64] The point of the nib was frequently made oblique for better writing. This is advised in Tawḥīdē's *Kitābah* no. 8 (from Rosenthal, *op. cit.*, 13).

[65] the knuckle.

[66] *maḥat* = to scrape.

[67] *sikīn al-barī*.

oldest. Its middle is thinner than its upper part because if it is according to what I have described, then the cutter can cut the pen and scrape it with great care using the knife's middle. If it is other than that, the cut of the pen results in a swollen middle section. After that, it is necessary to have another knife for the nib aside from the knife for the trimming, and the knife for scraping. This is better for the nib.

Description of the knife for the nib.[68] It is this knife which is the sharpest possible. The best is tempered in oil. If so, then the knife is not in danger of being nicked.

Description of that on which the nib is cut. It is essential that the thing on which the nib[69] is cut be of the best hard wood. It is not square on the sides and not hexagonal but it is round and smooth to make the cutting better. If it is square, then the knife does not cut in small pieces in the proper quantity and so makes it necessary for a second cut. In something like that, one fears its spoiling. If it is a hexagon, perhaps the knife would fall on the edge of the hexagon. Then the cutting would not come out properly. The round thing on which the nib is cut is easier for cutting and better.

Description of the inkwell. It is essential that the inkwell be of the best wood, of high price, and somewhat long. (7) The length must be the measure of a cubit[70] or a little less. The width of the inside must contain five pens for the writer, and for retaining luck seven pens, seven to rule the seven parts.

In the customary manner, the pens are nicely trimmed and cut in the way described. The length should be such that it can be held firmly. Raise the separation for the hand so that the top part can be extended on it from the pen. Know well the knife for trimming, the knife for nib cutting, and the stirrer of the inkwell.

The head of the inkwell is the place for the *līq*. It is round rather than square. The reason for this is that the square collects the ink in the corner which is upright and near the joining right angled sides. You cannot stir it; so the ink settles there. It stays there a long time, spoils, develops a foul odor, and changes color. Because of that, it changes whatever remains and what is close to it, i.e. the steeped ink. It is changed both in color and odor.

SECOND CHAPTER
ON THE PREPARATION OF SOOT INK

Description of Chinese ink[71] that resembles tannin ink. Take the best Fez ink[72] in the amount that you wish and pulverize it with yoghurt three days. As often as it dries, soak it with yoghurt and pulverize it. Then make it into sheets. It becomes like . . . and it sticks. It is wonderful.

Description of another ink that resembles tannin ink. Take *lāzward*,[73] tar soot, gum of scammony,[74] gum arabic,[75] and soot of sap of the pine,[76] of each one a part. Knead with water of the gum. It can then be used.

(8) Description of India ink. Take cow's butter and any oil—as much as the cow's butter, and the same amount of oil of the ben-nut tree,[77] clove,[78] violet, the <u>*laqat*</u>,[79] and whatever oil it is. Then you put it in a vessel.

---

[68] *sikīn al-qaṭṭ*.

[69] *maqṭ*.

[70] *ḍirā*ᶜ.

[71] *midād ṣīnī*.

[72] *midād fāsī*.

[73] *lāzward*, lapis-lazuli. This stone was known to the early Sumerians (BuA, 383–385) and Egyptians (*ḫsbd* in middle Egyptian). In India, it was well known in Rasakalpa as *rajavarta* (Ray, 156). The Arabic translation of Dioscorides gives *lāzward* as *armāniyā* showing what was generally supposed to be the origin of this stone. Galen (XII:223) considered lazward as a mineral with detersive properties.

Lāzward is described in al-Rāzī's *Sirr al-asrār*. "There is only one type. It is a dark blue stone in which there is a little red and possessing shining golden eyes." The latter, of course, refers to tiny spangles of included pyrites *Cf.* Ruska, J., Al-Rāzī's Buch Geheimnis der Geheimnisse, *Quellen u. Studien z. Gesch. d. Naturwissenschaften und der Medizin* **6**: 86, Berlin, 1937. In Boetius, lapis lazuli is differentiated from *lapis Armenius*. Hiller, J. E., Die mineralogie Anselmus Boetius de Boodt, *Quellen u. Studien z. Gesch. d. Naturwissenschaften und der Medizin* **8**: 137–139, Berlin, 1941. Boetius says that lapis lazuli is κυανὸς λίθος in Greek, in Latin, *ceruleus lapis*. He made use of many old sources in his mineralogy.

[74] *dukhān al-sāqmūnīyā*. This is the gum resin obtained by incision from the root of a plant found in China and Syria. It had an important place in the Muslim pharmacopoeia as a cathartic (Ainslie **1**: 386–389). Celsus (III: 20, 6) used scammony in cases of lumbrices. Resin of scammony was known in ancient Mesopotamia as coming from milky juice of the root. It is greenish-gray or brownish-green. It was used as a purgative (*cf. DAB*, 14 ff.). Diosc. (IV: 170) prescribed it in cataplasms with flour or on a pessary of wool in the uterus to kill the embryo, or leprosy and headache.

[75] *ṣamr* ᶜ*arabī*. This is the gum of the *ṭalḥ*. In Egypt the latter is the *Acacia sperocarpa* Hochst = *Mimosa gummifera* Forsk. In southern Morocco, it is the *Acacia gummifera* Willd.; in the Orient, it is the *Acacia vera* Willd. = *Mimosa nilotica* L. = *A. arabica* Willd. (*Tuḥfat al-aḥbāb*, 296.)

[76] *ṣanawbar* = pine. The pine was very common in ancient Mesopotamia (*ⁱˢLI* or *ⁱˢLI.PAR* in Sum. and *burashu* or *sikhu* in Akk., Levey, 123; *DAB*, 258). The *burashu* was used for incense and drugs. Turpentine from the pine was used as an expectorant, for the eyes, feet, breast, lungs, anus, and swellings. The Arabic word is a translation from the Greek πεύκη (*Iliad*-11: 494). It is found in Theoph. (III: 9) who knew the distinction between the male and female pines. It is also known in Diosc. (I: 69) and Serapion (438). It was thus well known to the Muslims viz. ᶜAbd. al-Razzāq (320) and in I. B. (1417, 1581) and in the *Tuḥfat al-aḥbāb* (298).

[77] *bān*. *Cf.* al-Ghāfiqī, 254–256. It is *Moringa arabica* Pers. according to Leclerc in I. B. (226). It may also be *M. pterygosperima* Gaertn. (Bonnet, E., Essai d'identification des plantes médicinales mentionnées par Dioscoride . . ., *Janus* **8**: 283–284.) Diosc. (IV: 157) claimed that it resembled tamarisk. Galen (XI: 845) used *bān* for skin diseases. In Greek, it is called βάλανος μυρεψική or in Latin *glans unguentaria* or *glandulae aromaticae*.

[78] *khīrī*. This is a type of clove. *khīrī* is also the Persian name. Maim. (294) states that the Egyptians call it *al-manthūr*. It is also called *al-khuzāma* and *al-bābūna*. The Arabs used this as an antispasmodic and diuretic (*Tuḥfat al-aḥbāb*, 422). I. B. (837) quotes al-Ghāfiqī who used it with vinegar for bad teeth.

[79] *laqat*. Perhaps oil of the gleanings of a grain plant or fruit (sometimes dates) fallen from a tree. (*Cf.* Lane, Dozy.)

Over it put another vessel.[80] Light a fire under the former vessel which contains oil and fat or whichever oil you wish until it becomes like vapor, all of it oily, and has risen into the upper vessel. There is a sublimate on the under side of the cover. It is gathered and used with this oil as the first ink was used. This black substance is good for dyeing the hair black.

Description of another India ink. Take two parts of cedar[81] or dried fruit of the pine,[82] or of them together. Put it into a new clay vessel[83] and put it into an oven[84] until it becomes charcoal.[85] It is taken out the next morning and pulverized a day on a stone.[86] It is soaked with water of cooked myrtle[87] and a little of vitriol[88] made according to the mentioned description. It is completely pulverized with water of myrtle, and pulverized with water of gum in such quantity that there is for every *raṭl* of pulverized charcoal ink two ounces of water of gum. If a little more is added, it is not harmed. If it is too hard, take the gum from it and knead it. Put it in layers and leave it in the shade. It comes out beautifully.

Description of Kufic ink. Take the rind of pomegranates[89] and procure wood to burn it. Take the ash and knead it with yoghurt[90] and a little of the moistened gum. Then make it into cakes and dry it in the shade. This is then the best type of ink.

Description of another Kufic ink. Take Greek gallnuts[91] and burn them until they become charcoal. Then pulverize it with water of the summer gum. (**9**) Make it into cakes and dry it in the shade. It comes out well.

Description of another Kufic ink. Take what you wish of the seed of dates.[92] Then put it in a vessel[93] and

---

[80] *anā'*

[81] *abzar*. From the Persian name of a mountain near Hamadan, in Persia, about 150 leagues west of Isfahan (Steingass) where the tree grows.

[82] *tamr al-ṣanawbar al-yābas*. *Ṣanawbar* = fruit of the pine. It is well known in Syria.

[83] *jarrah*.

[84] *furn* = a baking oven.

[85] *faḥm*.

[86] *ṣalābah*.

[87] *ās*. *Myrtus communis* L. In ancient Mesopotamia, *a-su-um* was employed as an aromatic astringent. An infusion of the berries, *ḥabb al-ās*, was used for leucorrhea and prolapsis of the uterus. In powder form, it was used for eczema, wounds, and ulcers, (*DAB*, 301). Myrtle was also used in enemas and in fumigation. In approximately the 21st century B.C., it was well known in medicine (Levey, 128, 150). The oldest medical text known, in the Sumerian cuneiform script, gives a prescription which included the skin of a water snake, *amamashdubkaskal* plant, pulverized alkali, barley, powdered fir resin, and root of myrtle (Sum. *GIS-GÍR*). All this is boiled together, then the liquid is decanted. Diosc. (I: 112) mentions it, μυρσίνη, to dye the hair, for eye inflammations, for erisypelas, and ulcers. *Cf.* I. B. (69) and al-Razzāq (11) for similar uses in the Arabic period. The *Tuḥfat al-aḥbāb* (11) says that it is also called *al-raiḥān* (also in al-Razzāq, 11). Synonyms are *mersīn* in Turk., *mūrd* in Pers. (al-Ghāfiqī, 9).

[88] *zāj*. According to al-Rāzī (10th century), there are five vitriols, black vitriol, *qalqadīs*, *qalqaṭar*, *sūrī*, and *qalqant*. (*Cf.* al-Rāzī 84.) These are described elsewhere in the same text (pp. 87–88) as "alum (*shabb*)," "*qalqadīs*, which is white vitriol, *qalqant*, which is green vitriol, and *sūrīn*, which is red vitriol." There is a confusion of terms for vitriol in al-Rāzī and other Arabic writers. The term *zāj* comes from the Persian *zāg* (*vide* Lane) (Lat. *atramentum* or *atramentum sutorium*).

The white vitriol (*al-zāj al-abyad*) is probably the double sulphate of aluminium and potassium. It was well known to the Greeks (as στυπτηρία) and to the ancient Egyptians. It is the Arabic *shabb*. Today $ZnSO_4 \cdot 7H_2O$ is known as white vitriol.

Qalqadīs (χαλκῖτις) is prepared according to al-Rāzī as follows: "Take white pure alum. Dissolve and purify it. Distill vitriol and verdigris. Mix them with water of the purified alum and leave it in a beaker."

Qalqant (χάλκανθον) is cupric sulphate or blue vitriol. This was frequently confused with green vitriol, ferrous sulphate. This was probably due to the fact that the vitriols, as was the case with almost all other chemicals used in antiquity, were impure. The methods of preparation as given by al-Rāzī (88) attest to this. For example, *qalqant* was prepared as follows: "Dissolve vitriol in water. Purify it. Throw on it copper filings and heat it until it is green. Purify it. Put it in a copper vessel. Dissolve it after you have put a half dirham of sal-ammoniac into ten dirhams of it.

"It is better when vitriol is dissolved that it be purified, put into a copper vessel, and dissolved after a half dirham of ammonium chloride has been put in ten dirhams of it until solid.

"Another type. Take yellow vitriol, heat it, and purify it. Add the same quantities of verdigris and vitriol. Leave it some days until it is dissolved and is green. Purify it. Let it become solid.

"To prepare *sūrīn* take vitriol, heat it, and purify it. Add the same quantity of iron-saffron. Cook it vigorously and purify. It comes out red."

Qalqaṭār, or colcothar in Paracelsus, is probably not a vitriol but is the product of the calcination of blue vitriol, or perhaps a peroxide of iron. Finally, the colors given may indicate the impurities and not the major substances themselves (Maim., 140, *Tuḥfat al-aḥbāb*, 144).

Throughout the text, *zāj* is probably meant to be green vitriol. *Cf.* also Diosc. (III: 80) and Galen (XII: 238).

[89] *ramān*. It is found in Galen, Rufus, Serapion, ibn Sīnā, and in other materia media as a stomachic, for biliary humours, and for other ailments. (See Maim., 75; Levey, 51, 108, 112). The rind of pomegranate is an astringent. Its pharmacological uses were many. It was used in ancient Mesopotamia also in the dyeing of purple with Murex and iron salts. Thompson, R. C., *Jour. Royal Asiatic Society* 81, 1934; Layard, A. H., *Nineveh and its remains* **2**: 98, London, 1849. Two main species of pomegranate (Sum. *NU. ÚR. MA*, Akk. *nurmū*) were known. The more common was *Punica granatum* L. It was called *rimmōn* in Hebrew. Diosc. (I: 110) gives the Greek as ῥόα which comes from the Indo-Eur. *sreu*. I. B. (1058) gives essentially the same uses. (*Cf.* also Carnoy, 231.)

[90] *laban ḥalib*. Yoghurt, an important milk product in a hot climate, has been known all over the Near East for thousands of years. It has adhesive properties.

[91] *ᶜafṣ rūmī* Greek gallnut. It was well known in Babylonian industry as shoemaker's gall and also in medicine (*DAB*, 205, 272). In tanning, it occupied an important place in the manufacture of good leather for religious purposes (Levey, 70, 71, 79, 112). *Cf.* Serapion (210), al-Razzāq (655), I. B. (1564), and Maim (295); *ᶜafṣ* is the κηκίς of Diosc. (I: 107). It was used as a febrifuge and intestinal astringent by the Arabs. As to *Quercus infectoria* A. D.C., its galls contain up to 60 per cent tannic acid. The drug *rāmik* (from Persian) was made from gallnut and aromatic drugs as mastic (*Tuḥfat al-aḥbāb*, 360).

[92] *tamr*. Its seed is an astringent used for serious ulcers and in collyrium for eye ulcers. *Cf.* I. B. (2241 *bis*).

[93] *qillah*.

lute its mouth. Put it in a warm furnace[94] a day and a night until it is burned. Then take it out. When it is cooled, open the vessel and take out the seed which has become like ash.[95] It is well pulverized and sieved[96] with the thick burned material. Then gum is taken and kneaded with it twice every day. It is made into cakes and then dried in the shade.

Description of Persian ink. Take the seed of the date that has been ripened in vinegar. Put it in a clay vessel. Take as much as you wish. Lute[97] the vessel with clay of the art.[98] The luting is done after a cloth has been put over the mouth. It is set down until it is dried a little. Then, if it is desired, the firewood is lit. It is shaken[99] from morning to night. If desired, it is introduced into the furnace for the two kinds of glass.[100] When it is taken out of the fire, it is set down until it is cold. Then it comes out black like charcoal. It is then made into cakes as desired.

Description of Iraqi ink. Anemones[101] are taken and stuffed into thin vessels[102] and buried in the dung[103] of asses until melted, watery, and dissolved. Then paper sheets are burned. What has been burned is gathered with the liquid and removed to dry in the shade. Then a dirham of it is taken, a dirham of gum arabic, and one-half dirham of pulverized gallnut. These are pulverized together with white of the egg. It is made into a ball. Then it is dried as has been mentioned, **(10)** put into the inkwell as needed with water of sorrel.[104] This is the best water for it.

Description of Nafuran ink. A large apparatus is constructed without holes or openings. Its middle is made as a square shelf. On the shelf is placed sandarac[105] and barley.[106] Then the fire in it is lit. The opening of the vessel is stoppered. It is left until all of it is burned. It is left to grow cold. The door is opened and the soot is gathered with leather sieves.[107] This is the leather that is not utilized for parchment[108] used in writing. This is the sieve of the miller.[109] Then it is put into a pot.[110] Water is poured over it and it is set over a fire. If dissolved, then it becomes a liquid like the acacia.[111] This would be the gum of the acacia.[112] When it is ripened, a bit of vinegar is poured on it and left until it is completed. Then a stone is smeared with water of camphor[113] and it is spread out on it until

[94] *atūn.*

[95] *ramād. Cf.* I. B. (1061).

[96] *nakhala.*

[97] *ṭān* = to lute with clay.

[98] *ṭin al-ḥikmah* = clay of the art. It was commonly used by Arabic alchemists and chemists to make apparatus air- and water-tight. Al-Kindī mentioned it frequently in his book on the preparation of perfumes (al-Kindī 27, 33, 34, 35, 39, 82, 83); in fact, this clay was used at times to cover an entire vessel before placing it in the oven. In one formula al-Kindī kneaded the clay with dung, hair and wine. Most frequently is this clay to be found in texts dealing with distillation where the alembic and ambix must be tightly sealed together before the operation. *Cf.* al-Rāzī (61).

[99] *ḥarak* = to shake or to stir.

[100] This may be a special type of glass furnace.

[101] *shaqā'iq* (sing. *shaqīq*)-sometimes the name is made more definite as *shaqā'iq al-nuᶜmān* (Diosc. II: 176). It is probably *Anemone coronaria* L. which is widespread in Palestine and Syria. Theophrastus (VI: 8; VII: 7) mentioned it as did Serapion (277). Its important uses as a drug are described in I. B. (1329), and in *Tuḥfat al-aḥbāb* (441). *Cf.* also al-Razzāq (106, 941) and Loew (111, 118). Other Arabic names are *saqir, laᶜib, laᶜīb, ḥannun al-daulah,* and *šaqā'iq an-naᶜmān al-mukhnath.* In Akk. it is *šamrattutu* and *šamartitu,* cognate to the Syriac *rᵉthīthā* = trembling. *Cf. DAB* (141). In Babylonia and among the Arabs it was used for head pain, diseases of the uterus, stoppage of milk in women, and dysmenorrhea.

[102] *qawārīr diqāq.* This vessel was not meant to be heated. It was thin enough to conduct the surrounding heat to the solution.

[103] *sirjīn.* The spontaneous oxidation of the surrounding dung gave off enough heat for the reaction desired. This was one way of achieving a temperature within a certain range before the thermometer was invented. The most common method which had been in use as early as the fourth millennium B.C. in Sumer was by using different types of ovens and furnaces. The early chemists were well aware that these were capable of yielding different degrees of heat. *Cf.* Levey (20–29) for a description of early heating apparatus and their developmental importance in aiding the growth of chemistry.

[104] *silq.* sorrel or beet. This word is derived from the Greek σικελωτικός, Latin *siculus,* "Sicilian." *Cf.* Theophrastus (VII: 1–6) and Diosc. (II: 114) for allied plants. In Akk., it is *silqa ŠAR. Cf. Tuḥfat al-aḥbāb* (171, 377, 397); al-Razzāq (142, 313); Loew (1: 358–360) Serapion (273). In I. B. (698), under *ḥummad,* λάπαθον, it is given as sorrel (*oseille*) or dockweed (patience). It was known by Galen (XII: 56) ibn Māsawaih, and ibn Sīnā. *Cf. DAB* (51) and Levey (87).

[105] *sandarūs.* I. B. (1238) says that it was used by ibn Sīnā and others as an emmenagogue, a diuretic, for the eyes, for fumigation, and internal hemorrhages. *Cf.* al-Kindī (104) and recipes 13, 14, and 64; al-Razzāq (821), al-Rāzī (41).

[106] *shaᶜīr.* This was the most common cereal grain in Babylonia. In the texts it occurs fifty times more often than wheat or emmer. It was used medicinally for poultices and as a stomachic (D, 99–101). Barley was also used in Babylonia in tanning solutions (Levey, 109). It was known in materia medica also to the Greeks (Diosc. II: 86). For Muslim usage *cf.* I. B. (1321, 1322), *Tuḥfat al-aḥbāb* (386) and Maim. (270) when it was known also as *sult* from the Akk. *siltu. Shaᶜīr* is also an old Semitic word.

[107] *manākhil al-julūd.*

[108] *ruqūq.*

[109] *manākhil qārūṭ.*

[110] *qidr* is a pot which can be heated over a fire.

[111] *qāqiyā* or *aqāqiyā* is the juice of the acacia. It is noted in Diosc. (I: 101) as ἀκακία, in Theophrastus (I: 2) as ἄκανθα. Serapion (6) knew it as did I. B. (1735). *Cf.* Maim. (12). In Babylonia, acacia bark was used in tanning (Levey, 112). *Cf.* also *DAB* (242). Ibn Baitar considered *qāqiyā* as the gum extracted from the *qaraẓ.* (*Cf.* I. B., 1758).

[112] *ṣamr al-qarḍ. Cf.* I. B. (1758, 1735), Maim. (278, 12). By *qarḍ* is usually meant the fruit of the acacia. *Cf.* Theoph. (IV: 2, 8), Diosc. (III: 13), *Tuḥfat al-aḥbāb* (46), Serapion (6), al-Razzāq (19). It is an Egyptian tree (M. E. *šndt*) *Acacia arabica* Willd. *var. nilotica* Del. Other species are to be found in the Sudan. It is an astringent and has been used so to the present day.

[113] *kāfūr* also called *gāfūr.* Obtained from *Cinnamomum camphora* Nees (Jabir IV: 108*a*, 108*b*). See al-Kindī (242–246). Leclerc (I. B., 1868) states that this tree is grown in India and China. The best are the *qaiṣurī* and *riyaḥī.* (*Cf.* Sontheimer III: 130.) The baros camphor from the east coast of Sumatra was the first brought to the West. Some also probably came from Borneo. *Cf.* Maim. (206). For a long time, camphor has been in use in India by the native practioners who prescribe it externally

both are dry. Then as many layers as desired are made. It is wonderful.

A special ink essence is made for the king from soot of refined storax,[114] soot of the sandarac, and soot of laudanum[115]—either together or separate. Its soot is a very strong black. Another ink is made from soot of bitumen[116] and also from soot of sulphur.[117] If it is desired that the *līq* in the inkwell not be spoiled and there not be a bad odor, then take the ink and put it in a vessel. Then enough clear water is poured over it to cover it. It is then filtered from its solution. Its water is changed three days. It is then put into the mortar and sorrel water poured on it, yoghurt or a bit of table salt,[118] and gum arabic. Then it is beaten in a mortar[119] until it has the consistency of glue. It is then put aside until needed. If it is desired to write with it some of it is dissolved in water.

THIRD CHAPTER

ON THE PREPARATION OF BLACK GALLNUT INKS

(11) Preparation of black shining ink. Ten parts of gallnut are taken and pressed. On it is poured six of the same water. It is then cooked until a sixth of the gallnut solution has disappeared. It is then purified. It is cooked in one-sixth of its weight of gum arabic, then boiled in a mild fire until one-third has disappeared. It is then brought down from the fire and cooled. One can write with it.

Description of another ink. An ounce of acacia gallnut is taken and pressed[120] with an ounce of gum arabic. They are mixed and there is poured on it a measure of water equal to eight times it. It is put in a flask[121] in the sun for three days. It is filtered after that. Four dirhams of Greek vitriol is put into it, and an ounce of the Iraqi if the Greek cannot be found. If it is summertime, it is left in the sun four days. If it is in the winter, it is left twelve days. It can be used to write with.

Description of another shiny ink. Two parts of gallnut are taken and pressed. On one part pour six parts of water and on the other two parts of sweet water. On the first another six parts of water are poured. It is soaked a day and a night. The two are gathered together in a new pot and cooked until its fourth disappears or its third. It is then taken down from the fire. It is purified. There is taken for it two ounces of scrapings[122] of gold. (12) It is pulverized and sieved, then sprinkled on. It is returned to the fire until it boils, then removed from the fire and purified. Two ounces of pulverized gum arabic is sprinkled on it while hot until it is nicely melted. Then it is put in a glass pot and used.

Description of instant ink.[123] Gallnut of the tere-

---

for sprains and rheumatism. When given internally, it is supposed to have the power of shortening the cold stage of intermittent fever. It is also supposed to be useful as a stimulant in typhus fever (Ainslie **1**: 48). Camphor is found in the Vagbhata of the ayurvedic period (A.D. ninth century) in a medical recipe. *Cf.* Ray, P., *History of chemistry in ancient and medieval India*, 71, Calcutta, 1956. Camphor was also used medicinally in ancient Babylonia (*DAB*, 80). Various camphors are in use today so that the confusion of the ancients lingers on. Japan or laurel camphor (from *Cinnamomum camphora* a tree flourishing in Japan, Formosa, and Central China) is in widespread use today. Camphor occurs in a variety of essential oils such as lavender, rosemary, sage, and others. There are also the Borneo, mint thyme, and buchu camphors. Camphors are today extracted by steam distillation of the botanicals, then separated usually by fractional distillation. *Cf. Aqrābādhīn* (110*b*) for use for the throat.

[114] *maïᶜa*. This was known in ancient Egypt and Babylonia. (*Cf.* Levey, 20; *DAB*, 321, 337, 340.) The Greeks (Theoph. IX: 7, 3; Diosc. I: 66) knew it well as στύραξ. The latter name comes from the Syriac *aṣṭūrkā*. The Arabs distinguished between dry storax (from *Styrax officinale* L.) and liquid storax (from *Liquidambar orientalis* Mill.) *Cf.* Steuer, O., *Myrrhe und Stakte*, Wien, 1933; *DAB*, 335. *Vide Tuḥfat al-aḥbāb* (58, 238); Ducros (226–228); I. B. (97, 2196); al-Razzāq (513, 529); al-Kindī (5, 63, 65). It was used as an aromatic for thousands of years.

[115] *lādan* or *lādhan*. It is the resin from the plant, *Cistus ladaniferus* L., κίσθος, and other species. It was known also as λάδανον to Diosc. (I: 97). It is to be found in al-Razzāq (504), al-Kindī (60), *Tuḥfat al-aḥbāb* (241), and Maim. (208). It is still used today as an astringent, antidysenteric, and in collyriums. *Cf.* *DAB*, 335, 344. *Cf.* also I. B. (1999) and Pliny (XII: 37). The origin of *lādanon* is Semitic (Carnoy, 156).

[116] *zift*. There is general confusion in the Greek (Diosc. I: 72) and Arabic literature in regard to the bitumens, pitch, and asphalts. This is discussed in the *Tuḥfat al-aḥbāb* (150). *Cf.* Ducros (178) and al-Razzāq (276, 758). *Cf.* Maim. (102, 138, 168); al-Kindī (135); I. B. (1114). *Zift* was used in many medicinal preparations as well as in the building of liquid containers, especially cisterns, and on streets as a sealant. For its uses in ancient Babylonia, *cf.* Levey (168), *DAB* (239), and Forbes, R. J., *Bitumen and petroleum in antiquity*, Leiden, 1936. In India (Ray, 29, 170) bitumen was probably obtained from Baluchistan and imported to the Indus Valley 5000 years ago.

[117] *kibrīt*. Sulphur was well known in ancient Mesopotamia. *Cf.* Levey (39, 127, 128) for its preparation and uses in Babylonian pharmacology. Aristotle, Galen, and al-Rāzī among others knew it. Diosc. (V) discussed it as a simple. Sulphur was employed in ancient times a great deal for ulcers. It was used much in 18th-century India and earlier for itch and cutaneous infections. Hippocrates and Celsus also used sulphur. *Cf.* Ainslie (**1**: 411–414); Ray, *passim*. In Alaune (41) sulphur is described. Its nature is described as like the nature of the arsenics. When it was added to mercury and then distilled, the author obtained cinnabar. In the same way, when tested with copper, a burned copper (oxide) was obtained. "What is desirable is that its burning and its oiliness and whiteness disappear as with arsenic." A red (?) sulphur is mentioned. Sulphur is discussed in I. B. (1880).

[118] *milḥ al-taᶜām*. Many substances were called salts by the Arabs (Maim., 221). Al-Kindī (71*b*) used salt frequently in his recipes. Salt is discussed in I. B. (2164), al-Razzāq (542), Diosc. (V: 109), and Serapion (358) among others. Salts are also discussed in Alaune (51 ff., 80–81). The table salt used in the mediaeval period in most places was obtained largely from the sea. That which was mined was more expensive because of the transportation costs. Salt was generally impure.

[119] *hāwun*. *Cf.* Levey (13) for Babylonian mortars. Many types of mortars and terms relating to them were known in the lexical texts.

[120] *raḍḍ*.

[121] *qinīnīah*.

[122] *qilf*. Pers. *galf*. Possibly the coloration on the surface of impure gold exposed to the atmosphere.

[123] *ḥibr sā'atah*.

binth[124]—yellow green, is taken, Greek vitriol, and gum arabic—of each a *mithqāl*. It is all pulverized and put into a vessel which is wide-mouthed. Two ounces of salt water are poured on it. It is well beaten. It can be used for writing immediately on paper and parchment. This is the description of the Iraqi way.

Description of a black ink. Three ounces of gallnut are taken, an ounce of glass,[125] and one and a half ounces of gum. The gallnut is pressed. On every part of it, eight parts of water are thrown. It is soaked in it a day and a night; if more than that, it is better. Then it is put on a low fire until a third has disappeared. When the gallnut has deteriorated, then it is well cooked. The gum is dissolved in water before the cooking of the gallnut; it is completely covered until all of it becomes viscous like honey. When the gallnut is cooked, the gum is thrown on it. It is left a little while until all of it is dissolved in it. After it is pulverized, the vitriol is put on it. If it is not enough, put more on it. Never throw the gum on it unless it has been soaked.

Preparation of a dry ink. Green gum is well pulverized until it becomes like collyrium. A part of it is taken, and a part of the gum arabic. The gum arabic is dissolved with water. One half part of vitriol is taken. (**13**) All of it is gathered with glair of egg, gallnut, and gum arabic until it becomes like dough. It is made into a ball and put into a vessel. Be sure that no wind and dust reach it. It remains for a long time. When it is needed, it is put into a vessel. Drip water[126] on it from another vessel in a needed amount until it is dissolved. Then one can write with it.

Preparation of ink for the common people. Green gallnut is taken and pressed in quarters and thirds. It is put in a narrow mouthed flask.[127] Then water is poured on it. It is then placed on the fire. A low fire is used. When half of it disappears, it is clarified. To write with it, for every *raṭl* of water of the powdered gallnut use five parts of the solution prepared and one half ounce of green vitriol.[128] Write with it.

Preparation of ink of the myrobalan.[129] Yellow myrobalan is taken and pressed with seed. It is put into a thin vessel after it has been measured. Then pour on it two-thirds its weight of water. It is put in the warm sun for four days. Then it is purified and gum arabic is put into it. It is returned to the sun and left until dissolved. Then a little water of yellow vitriol is added and a bit of pulverized green vitriol. It is stirred and then used to write with.

Preparation of a sunny ink without use of fire. Ten dirhams of gum arabic are taken, six dirhams of green gallnut without worms, and four dirhams of very shiny Cypriote vitriol.[130] (**14**) Everything of the mixture is powderized and sieved with a thick sieving cloth. It is weighed after the sieving so that it is not lessened. On it is poured one hundred dirhams of clear water. It is stirred with the finger until the gum is dissolved. Write with it immediately.

Preparation of a foreign ink. Four *raṭls* of pure water are taken and put in a pot. Also four ounces of gum arabic, the same amount of gallnut, and the same amount of tamarisk[131] gall are taken and each one pulverized. The gallnut and tamarisk gall are thrown into the water and cooked until half has disappeared. This is reckoned by means of a stick. When it comes to the half, the gum is thrown into it. One and a half ounces of *lukk*[132] are thrown into it in powdered form. When it has boiled twice or thrice, it is taken down and left until it is settled. When it has settled and cleared, the clear part is taken. It is the best ink. The residue[133] is taken and put into the inkwell. If it does not write and it is burned, then its gallnut is pulverized and

---

[124] *ᶜafṣ al-buṭm*. Terebinth was used as a detersive substance in Babylonia (Levey, 124). *Cf. DAB* (253). The butum tree is usually the *Pistacia Terebinthus* L.; Sum. *ᶦˢLAM.GAL*—Akk. *buṭuttu* = Syr. *beṭmᵉtha* = Heb. *bōṭem*. It was used medically for rheumatism and local pains. Hippocrates used the fruit, bud, and resin. *Cf.* Theoph. (III–V), Diosc. (I: 71), *Tuḥfat al-aḥbāb* (178), Maim. (66), Jābir (VI: 191*a*).

[125] When dry, the glass gives a sparkling appearance to ink.

[126] From *qaṭar*.

[127] *qamqam*. This is a flask which can be heated. It may have a narrow mouth.

[128] *zāj akhḍar*. Green vitriol = vitriol of Cyprus (*al-zāj al-qubruṣī*). *Cf.* Ducros (133). It was unknown to Dioscorides. See also I. B. (1080), Maim. (140). It was well known to the Babylonians (*DAB*, 169 ff.) who worked in metal and leather. The vitriol of the leather worker was called *ˢᵃᵐkamme ashkapi*. In Pliny, it is *atramentum sutorium*. In Syriac, it is *ekam*. *Cf.* Ray (173) for *valuka-kasisa*, sulphate of iron, in *Rasaratnasamuchchaya*. In Ducros (133), *zāj akhḍar* is *shaḥirah* in present day drug commerce. The latter term is also one used by al-Jildakī in his *Kitāb al-burhan fī asrār ᶜilm al-mizān*. *Cf.* pp. 37, 38 in ibn Sinā.

[129] *halīlaj* or *ihlīlaj*. Bayān (19) used myrobalans as aphrodisiacs. *Cf. Tuḥfat al-aḥbāb* (43, 122, 126). *Cf.* al-Razzāq (253). *Cf.* also Serapion (71, 226), Ducros (13–15), I. B. (2261), and Maim. (112). Myrobalan juice is used also in coloring of leather in this text (pp. 75, 76). For the use of chebulic and emblic myrobalan in medicine, see Ainslie (**1**: 237–241). It is an astringent and frequently used as a cathartic. *Cf.* al-Kindī (4) for a full discussion of myrobalans. See note 301.

[130] *zāj qubruṣī*. green vitriol, probably the purest in antiquity.

[131] *'adbah* or *'adhbah*. Many uses for tamarisk were well known in Babylonia (*DAB*, 279, Levey, 123). Equivalent probably to *athl* and known to Diosc. (I: 89) as ἀκακαλλίς and Theoph. (V: 4, 8,) as μυρίκη. It is discussed in I. B. (17) and Serapion (43). Ducros (56) says that it is used today as an antidysenteric, astringent, and dentifrice. It is usually *Tamarix orientalis* Forsk. More frequently the gall is called the *tamr al-athl*, in Pers. *kazmāzak*, and in Berber *tākkawt* (*Tuḥfat al-aḥbāb*, 23, 106, 228). *Cf.* Maim. (9). Frequently, the Arabs confused the tamarisk with its fruit in their terminology. Leiden Or. 576 (181*a*) gives gum of the terebinth as resin of the Nabataeans useful for sores, coughing, for encouraging urination and removing the secundine.

[132] In Persian *lāk* (Steingass) but originally Indian. Lac, at present, is a product of the East Indies especially Bengal, Pegu, Siam, and Assam, and is produced on a number of trees. *Cf. Tuḥfat al-aḥbāb* (245), Dozy. Ducros (210) states that it is today sold in the bazaars as a tonic and an astringent. *Lukk* comes from Hind. *lākh* which is from Sanskr. *lakṣa* meaning a mark or sign.

[133] *tifl*. It may also mean precipitate.

soaked in water for three days. Its clear water is taken and spread on the bottom.

Preparation of dry ink for travel. The best green gallnut is taken and well pulverized like a collyrium paste. Also pulverized for it is the same amount of gum arabic. Then half its weight is taken of green vitriol. It is well pulverized and all of it gathered in the white of one or two eggs[134] until it becomes like dough. Then it is formed into a ball and put into a vessel and stoppered at the head to prevent the access of air or dust. It is allowed to stand upright a long time.

Preparation of another powdered dry ink. Gallnut, gum arabic, vitriol, and acacia in equal parts are taken. All are pulverized with water of fresh St. John's bread[135] until it is dried. (**15**) It is then removed and dissolved with the necessary amount of gum water when it is needed. Write with it.

Preparation of ink which is made with the water of the myrtle alone. Seed of old myrtle is taken and cleaned in water. For every *raṭl* of it, there are dissolved three *raṭls* of gallnut water and four ounces of the extract of the twig of the myrtle. Then it is put in the warm sun for seven days. It is squeezed and clarified. For every *raṭl* of this solution add one-half *raṭl* of gum arabic. It is set down for a day and a night until it is dissolved. Then there is added to it Cypriote green vitriol[136] in an amount which is enough. If it is worked with Egyptian vitriol,[137] it gives better results. It is clarified and used to write with.

Preparation of ink with water of the Syrian mulberry.[138] The water is taken that flows from the Syrian mulberry and into it is thrown pulverized gum arabic and a little water of the green gallnut. The water of the green gallnut is not increased. It is burned and suspended in the shade. Every day a dirham of gum is thrown into it. This is done five times in five days with a *raṭl* of that water. Write with it.

(**16**) Preparation of ink for religious books. Gallnut is taken and pressed to the size of the chick pea.[139] It is then measured and put in a boiling vessel. On it are poured three parts of the same water. Then a fire is lit under it until it returns to two parts. It is cooled and clarified. In it is thrown green vitriol, whatever is needed and gum arabic—for every one of water—a part and a half of gum arabic. Then write. Some people who cook it do so until the water returns to two-thirds or one-third or as desired.

Preparation of ink for owners of religious books. A part is taken of pressed green gallnut and on it are poured five parts of water and cooked until it comes to one and a half parts or one part. It is then clarified and put into a vessel containing vitriol. Vitriol is taken and put into a pot. On it is poured the same amount of water. It is kept in the sun three or four days. Then a part of the water of the gallnut and a part of the vitriol are mixed. Before that, gum arabic had been taken and water poured over it and left in the sun a day or more until dissolved. Then two parts of it are mixed with two of water, stirred gently, and then used for writing. If it is desired that it be deep black, then one-half ounce of burned pulverized vitriol is added. It is used immediately. Write with it.

Preparation of a dry, black ink. An ounce of gallnut is taken. It is pulverized until it becomes like dust. Do the same with gum arabic. It is kept free from dust. It is then suitable for the best book.

Preparation of another ink. The water of the black, ripened Syrian mulberry is taken in the amount of a *raṭl*. With it are put ten dirhams of pulverized sieved gum arabic. A little vitriol is added to it. It is put into a pot in the sun forty days. It is used after that.

Preparation of ink from iron filings.[140] The gallnut is boiled with iron filings until one-third of the water disappears and two-thirds remains. It is then clarified in a pot and put in the sun for a day. For every *raṭl* of water, there is thrown on it a dirham of vitriol. (**17**) Add whatever gum is needed. It comes out wonderfully. If a wine color is desired, the gallnut is pressed[141] and soaked with the filings. To every *raṭl*, five *raṭls* of water are added. It is boiled[142] well and left. When cold it is clarified. To every *raṭl* of water, there are added four dirhams of vitriol. Write with it.

Description of another good ink. Gallnut is taken and broken up into quarters or smaller. On it is poured enough water to cover it. It is placed in the sun for two days. It is pressed and boiled on the fire. The necessary vitriol is added as well as gum.

Description of another ink for religious books. Gallnut is taken and pounded down to the size of a chick

[134] The white of eggs was frequently used by the Arabs for its adhesive property as well as for the final glazed appearance it helps to give the ink.

[135] *kharnūb*. Called fruit of the carob or St. John's bread. It was a staple of the materia medica (*DAB*, 186) in Babylonia, India (Ainslie **1**: 364–365), and in Arabic times (I. B. 762). This name probably comes from the Hebrew *ḥarrūb* or Aramaic *ḥarrūbā* (originally *kharubu* in Akk.) and refers to the *Ceratonia Siliqua* L. and its fruit, a tree native to Palestine. The Greeks knew it as κερωνία (Theoph. I: 11–13; IV: 2,4) and κεράτια (Diosc. I: 114). *Cf.* also Serapion (114), al-Razzāq (920). The *Tuḥfat al-aḥbāb* (204, 423) calls it *tarīlt* (?) in Berber; Maim. (392) calls it *tāslīghma* in Berber. Ducros (95) lists it as being sold in the bazaars as a laxative and cough reliever.

[136] *zāj akhḍar qubruṣi*.

[137] *zāj miṣr*. Al-Rāzī mentions the yellow Egyptian vitriol and the white. It is uncertain here which is meant. See al-Rāzī (87). It is probably an impure alum.

[138] *tūt shāmi*. In Babylonia (*DAB*, 316–317) the mulberry was well known as *ⁱˢMÌS MÁ.KAN.NA* in Sumerian and *musukkānu* in Akk. It was rare in their materia medica. Dioscorides (I: 126) gives *Morus nigra* L., μορέα, as the important mulberry. I. B. (434) gives *tūt* as being very useful in the old materia medica for the stomach and bile. *Tūt* is apparently Persian (Lane).

[139] *ḥimmaṣ*, the size of a chick pea. It was common in the Muslim world to relate the size of a thing to a seed or some other part of a common plant or animal.

[140] *burādat al-ḥadīd*.

[141] *raḍḍ*.

[142] *ghala*.

pea and smaller. It is put into a pot and on it is poured ten times its quantity of sweet water.[143] The fire is kept up until the solution returns to its half or third whichever is better. On it is thrown enough vitriol and the gum arabic in the amount needed. You write with it.

Preparation of another ink. One part of gallnut is taken, two parts of gum, and one part of vitriol. All of it is pulverized and covered with water. It ferments for a night. Then add water the next day until it comes to the amount necessary. Write with it.

Preparation of another good ink. A small gallnut without worms is taken in the weight of three ounces and soaked in water in a proper pot for four days. It is put on the fire and good green vitriol is thrown on it. It is left an hour after the fire has become strong. On it is thrown clean, pulverized gum arabic. It is left a night. (**18**) With the coming of daylight, it is purified and put in a glass vessel. Write with it. It is a good black.

Preparation of another ink. Gallnut and pomegranate rind are taken, pressed together, and soaked in some water for three days. Then blue vitriol is poured on it little by little while stirring until it is seen that it has become a strong black. If the *qalqant* is not available, then use its substitute, Persian vitriol. Gum arabic is added. It is then removed from the fire. It is good.

Preparation of an ink with which one writes in copy books: Thirty macerated gallnuts are taken. On them is poured three *raṭls* of water. Then it is cooked on a low fire until one-third has disappeared. It is then clarified and on it is thrown five dirhams of vitriol and nine dirhams of gum arabic. It is left in the sun for a day. If the black is not in it, then add vitriol to make it good.

Preparation of still another ink. There are taken three ounces of gallnut, one ounce of vitriol, and one and a half ounces of gum arabic. The gallnut is crushed and on it is thrown eight times its weight of sweet water. It is in the water a day and a night, the longer, the better. It is then placed on a low fire for a night until it is decreased by a third. The proof that it has been well cooked is that when the gallnut is squeezed it disintegrates. Then gum is soaked in a little of that water before it is cooked until it is the viscosity of honey. As much gum is then added as is on the fire. An equal amount of vitriol is added. (**19**) It is then taken down from the fire and clarified. Write with it.

FOURTH CHAPTER

ON THE PREPARATION OF COLORED INKS

Preparation of the red, yellow, and green inks. Twenty shekels[144] of sour pomegranate rind are taken. If it is moist,[145] it is better, else the dry type must be used. Also the rind of the green walnut[146] in the same amount, twenty *mithqals* of the green gallnut, twenty *mithqals* of Isfahan antimony,[147] and juice of the myrtle[148] equal to that amount. It is put in the sun for forty days, then clarified and placed in three other flasks. In one of these flasks, pulverized cinnabar[149] is put. It is stirred with the pen. This is red ink. Then pulverized verdigris[150] is put in another pot and stirred.

[143] Unsalted water. It is not sweetened water.

[144] *shaqal.*

[145] *raṭab.*

[146] *qishr al-jawz al-akhḍar.* Many nuts were used in ancient times according to Maim. (82). This is probably the unripened walnut. It is *jawz* in Persian, Steingass (377). It is the walnut in al-Kindī (32), *Juglans regia* L., whose oil is used to extract oil from cotton seed and others. In Greek, it was called καρύα βασιλικά (Diosc. I: 125) and was used for the stomach, cholera, as an antidote for poisons, and for inflammations. In Arabic times (*cf.* Maim., 82, *Tuḥfat al-aḥbāb*, 98–100), there was much confusion in the terminology for the various nuts. I. B. (525) states that the walnut was used by the Arabs for the stomach, ulcers, kidneys, and impetigo. It is still used in Arabic medicine (Hooper, 131) and also grown in Iran and Iraq.

[147] *ithmid iṣfahānī.* Antimony sulphide. According to Maimonides (Maim., 27) it is mined in the Maghrib and is called *kuḥl al-zurqa.* That found in the Orient is called *kuḥl isbahanī.* (the dry collyrium of Isfahan) which is equivalent to *ithmid iṣfahānī.* It was well known in antiquity. *Cf.* Diosc. (V: 84) and Galen (XII: 236). It is mentioned in al-Razzāq (20) and Zahrāwī (173). Ducros (197) refers to it as antimony or stibium. It is usually uncertain whether the sulphide or antimony is meant in the literature. *Cf.* Ducros (113) which states that a sample of ancient Egyptian *kuḥl* was analysed and found to be the sulphate of argentiferous lead. Antimony in Egypt and Babylonia had many uses in cosmetics and medicine (*DAB*, 44). In India (Ainslie **1**: 495–498) the sulphide is commonly used, knowledge of it having probably come from Arabic sources. *Cf.* Jābir (I: 54*a*). A description of *ithmid* is to be found in Aristotle's lapidary (51). *Cf.* also note 2 on page 175. The Isfahan variety was usually considered to be the best. *Cf.* al-Rāzī, 45, 151–152.

[148] *usārat al-ās.* The myrtle is also called *al-rayḥān* (*Tuḥfat al-aḥbāb*). It is *Myrtus communis* L. I. B. (69) uses *al-rayḥān.* *Marsīn* and *ḥimblās* are also used as synonyms. The myrtle was used pharmacologically in Babylonia (^{riq}*GIR* in Suml, *āsu* in Akk.). *Cf.* *DAB* (301), Levey (128, 150). Myrtle, according to I. B., was well known to Galen and Dioscorides (I: 112). It was used as an astringent, a stomachic, and for other pharmacological purposes. Ducros (4) says that myrtle abounds in the Mediterranean region and is sold in the bazaars as an astringent, for cataplasms on ulcers, and hemorrhoids. *Mirsīn* comes from the Greek μυρσίνη which comes from the word for Smyrna. The tree is indigenous to the Mediterranean area. It is still used also for leucorrhea and prolapsus of the uterus.

[149] *zanjafr.* Cinnabar had many uses in ancient Babylonia (*Cf.* *CAS*, *passim*). It was called ψωρικόν in Diosc. (V: 99). It is the sulphide of mercury used by Jābir (II: 54*a*; IV: 113*a*; VI: 185*b*) in his antidotes. Al-Rāzī (106–110) gives a full chemical account of cinnabar. *Tuḥfat al-aḥbāb* (147) says that it is "prepared from mercury (*zībaq*) and sulphur but it is at the same time a mineral." *Cf.* appendix of I. B. (1132) regarding preparation of cinnabar from sulphur and mercury in the tenth century.

[150] *zinjār.* Comes from the Persian *jangār.* Known in Babylonia and Egypt, it is in Diosc. (V: 79). Today *zinjār* is applied to natural verdigris. Al-Rāzī (51, 91) lists *zinjār* as a compounded material and not elementary "natural" matter. According to the lapidary of Aristotle (71), it is made from vinegar and copper. Then it is useful for sores, leprosy, burns, and fistulas. In ancient times, *zinjār* was not known in a pure state. I. B. (1131) gives its medical uses for ulcers, the nose, and in ophthalmology. It is still used today in eye ailments.

This is green ink. In another pot put pulverized yellow arsenic.[151] It is stirred. This is yellow ink. When it thickens in the flask, add this water.

Preparation of an ink for parchment especially to make it look like gold. Pure red arsenic which has not been mixed with anything is taken. It is well pulverized. Then pure good saffron[152] without oil[153] or fat is taken and wrapped in a clean cloth. It is put in pure water until the bundle is moistened. It is then squeezed on the arsenic and water of gum put on it. Write with it. It comes out like pure red gold.

Preparation of an ink for people of the sword. One part of gallnut is taken and broken up. On it is poured three parts of water. (**20**) The fire is lit until the solution returns to one part. Then green vitriol is taken and two parts of the water poured on it and stirred in the vessel for three days. Yellow myrobalan is pressed with its seeds. The seeds are not broken. On a part of it, there is poured three parts of the water. It is put on the fire until one part is left. It is good.

Preparation of a red ink. Gallnut is taken and broken up. From the inside the red and black are thrown away while the rind of the outside is left. It is soaked in water after all of it is washed with water. It is put in a vessel and stirred. When a clean foam appears, then it is left in its condition until it is dry. It is well pulverized until it becomes like dust. It is beaten with that water. It is left for an hour. Gum arabic is put into it. Write with it.

Preparation of an ink with which one can write on the same day. An ounce is taken of green whole gallnuts. It is well pulverized and sieved with a thick silk cloth. One ounce of the best Cypriote vitriol which has in it eyes of gold is added. It is pulverized and sieved again. White gum arabic and two ounces of the strongly shining red type are taken. It is pulverized and sieved. On the gum is poured a *raṭl* of water. It is squeezed with the hands until it is dissolved. Then the gallnut is thrown into it and also the vitriol. It is stirred until all is mixed. Then the redness is examined. If it is close to peacock red, then it is good. As much water is mixed with it as is possible. It is put in a glass vessel.[154] Write with it immediately.

Preparation of a ruby red ink. Saffron is taken and washed. (**21**) It is then pulverized until it becomes like a paste.[155] It is then beaten with water of the crushed[156] white gallnut as in the first process. It is left an hour, then beaten with water of dissolved gum arabic. It is then stirred vigorously and used.

Preparation of red ink. Green gallnut in halves or thirds is taken and crushed. For every measure of gallnut, nine of water are used. It is placed in the warm sun five or seven days. Then the water above the gallnut is removed with a fine cloth. Then for every ten dirhams of gallnut five or ten of gum arabic are taken. It is well pulverized. Seven dirhams of good vitriol are taken and the gum arabic poured first. When the gum is dissolved, then the vitriol is poured on it. It is stirred by hand, that is, with a pen. If the color on the pen becomes white, then not a thing is added. If vitriol is added, then it will burn.

Preparation of a peacock colored ink.[157] Yellow myrobalan is taken and soaked with its seed and cooked. One ounce of pure Greek vitriol is cooked with water of the myrobalan and one-half ounce of gum arabic. Write with it. It comes out beautifully.

Preparation of a peacock-blue ink[158] for parchment.[159]

---

[151] *zarnīkh*. Already known for thousands of years, Aristotle's lapidary (27) mentions three types, yellow, red, and dust gray. *Ibid.*, 162, notes 2, 3, gives as a Latin synonym *elzarmeth*. The "yellow arsenic" is actually auripigment or orpiment while the red is realgar. The red and yellow were the only sulphides known to the Arabs. White arsenic is actually As_2O_3. The best mediaeval description of sulphides of arsenic is given in Alaune taken from the Arabic and Latin texts. They are late texts but nevertheless valuable. *Cf.* pp. 84–88. Jābir is quoted in regard to the preparation. Described are the distillation and the washing. Only the red (*zarnīkh aḥmar*) and the yellow (*zarnīkh aṣfar*) are mentioned, thus giving a more accurate picture of the sulfides of arsenic. For orpiment in Babylonia, *cf.* Levey (96, 161). Dioscorides (V: 104) knew realgar and orpiment (V: 105), σανδαράκην and ἀρσενικόν respectively.

[152] *zaᶜfarān*. According to Maim. (135), it is also called *al-jādī* or ***kurkum***. The Akkadian $^{\check{s}am}$*azupirānu* is obviously the cognate of *zaᶜfarān*; $^{\check{s}am}$***kurkanū*** in Akk. is the cognate of ***kurkum***, ***karkom*** in Hebrew. Saffron was used in Babylonian medicine as an emmenagogue; in industrial technology it found use as a dye (Levey, 77, 105). The Sumerian $^{\check{s}am}$*Ḫ A R. S A G. Š A R* is equivalent to $^{\check{s}am}$*azupiru*. Saffron, *Crocus sativus* L. was frequently confused with turmeric, also a yellow dye. Saffron was also used as a condiment (*DAB*, 159–160). Dioscorides (I: 26) (κρόκος) prescribed saffron root as a diuretic. In the Indian materia medica saffron is prescribed in nervous affections unattended with vertigo, in melancholia, hysteric depressions, and typhus fever (Ainslie **1**: 354–357). Al-Kindī (130) used saffron in many perfumery recipes. (*Cf.* 383–384.) Many Muslims (I. B., 1110; *Tuḥfat al-aḥbāb*, 151) placed saffron among their cardiaca and hypnotica. Al-Rāzī and ibn Sīnā were aware of the misuse of saffron in affecting the appetite and causing nausea. Formerly grown in Isfahan, the saffron plant is now cultivated in regions of Pampur and Kashmir. The drug has a stimulative and antispasmodic action (Hooper, 107).

[153] *zait*. The olive tree was never grown successfully in ancient Mesopotamia (Levey, 87) until Sennacherib, who called the oil *šaman* ^{u}*sirdi* (oil of the fruit tree). Throughout ancient times, however, the sesame remained the most important source of oil. Even in Talmudic times, sesame oil and not olive oil was used in Babylonia. The latter is found in the materia medica of Dioscorides (I: 30–34). In the Arabic period, the leaves of the wild olive were used as a vulnerary, and the oil in laxative enemas (Ainslie **1**: 268–269). The olive tree is not cultivated in India. Al-Kindī (130*b*) used it in perfumery. I. B. (1141) discusses the use in cataracts and for the bite of a scorpion. It is, of course, very common today and used as a stomachic in the Near East.

[154] *qarūrah zajāj*. Usually spelled *qārūrah*. It is a flask with a long neck (Wehr, 671). *Zajāj* = glass.

[155] *marham*. Like a salve or plaster.

[156] *marḍūḍ* = *marḍuḍ*.

[157] *ḥibr ṭāwūs*.

[158] *ḥibr azraq ṭāwūs*.

[159] *raqq*.

Broken-up seed of coriander[160] is taken and cooked until it attains the viscosity of a paste. On it are thrown five dirhams of gum and a dirham of *lukk*. Write with it.

Preparation of a rose-colored ink. An ounce of red lead[161] is pulverized on a stone. (22) On it are thrown a dirham of natron[162] and two dirhams of gum arabic. It is rubbed with the hand until it is good. Write with it.

Preparation of a pistachio ink. Pomegranate-like cinnabar is boiled. It is then pulverized like a salve. Dissolved gum is beaten in water. Water of the dissolved *lukk* is beaten and all stirred vigorously and used.

Preparation of a purple ink. Gallnut is crushed. On it is thrown five times this amount of water. It is boiled, then taken down from the fire. When cold it is clarified. On every *raṭl* of it, five *raṭls* of arsenic[163] are thrown. Use it.

Description of another ink from anemones. When it is reddened, it is well pulverized. Then it is pulverized in wine vinegar[164] and put on the fire. Gum is added. Write with it.

Preparation of another ink from anemones; it is ink called *Barsān*.[164a] The stalks are removed from roses. Wine vinegar is put on it to cover it. It is boiled until its color comes out on the fire. It is taken down. To it are added a dirham of water of myrtle and the same amount of gum arabic. It is boiled twice until the water is much reddened. It becomes thick. Then one writes with it.

Preparation of a ruby-colored ink. The best pomegranate-like cinnabar is taken and powdered until it becomes like a paste. It is beaten with water of crushed white gallnut. It is left for an hour. Dissolved gum arabic is put in it. Write with it.

Preparation of grand basil[165] ink. Take the red purple material and gather of it one quarter of a *raṭl*. It is put in a mortar and pulverized until it is well mixed. It is clarified and put into a glass vessel. (23) *Lukk* is put into it. This is good and pure.

Preparation of another good ink. Three dirhams of indigo[166] are taken and pulverized on a stone with warm water until it becomes like paste. Then a dirham of verdigris is thrown on it. It is rubbed until its color is green and it is pretty. Write with it.

Preparation of another ink. Three dirhams of gum

[160] *kazburah* or *kasburah*. The coriander (*kusibirru* *ŠAR* in Akk.) is the *Coriandrum sativum* L. whose seed was used in medical recipes in Babylonia (*DAB*, 64, 93). In Nepal the coriander is common and is called *danga*. It was otherwise imported from Egypt or Syria to be used in their materia medica. Diosc. (III: 63), κορίαννον, used coriander seed for "creeping ulcers," and inflammation of the skin. I. B. (1933, 1926) gives the Arabic medical uses for the stomach, calming excessive erections, and other purposes. The juice of the coriander is lethal in doses of four ounces. In North Africa, the popular pronunciation is *quṣbūr*. Morocco is an important producer of coriander (*Tuḥfat al-aḥbāb*, 230). Ducros (199) states that coriander is sold in the bazaars of Egypt as a carminative, stomachic, condiment, and digestive.

[161] *sīlqūn* or *saliqūn*, in Pers. *asranj*. Sometimes in Ar., it is *usrunj zarqūn*. The latter's etymology is doubtful (Dozy, 225). The Castilian *azarcon* probably came from it. Perhaps it is from the Persian *adhargun* "color of fire." In Greek it is similarly συρικόν or in Diosc. (V: 88) σάνδυξ. It is also called minium. In Babylonia, it was used in dyeing (Levey, 138). In Arabic times, I. B. (74) lists its use for intestinal ulcers and as an unguent with oils. In the lapidary of Aristotle (181–182), minium is declared to come from lead (*raṣāṣ usrub*) which has changed to become red in the fire. In a salve, it is claimed, it heals wounds and is effective as a medicament in various ailments. *Cf.* al-Rāzī (50, 91, 226) for the properties of red lead. Jābir also used minium (II: 54*a*; VI: 185*b*). In the *Tuḥfat al-aḥbāb* (54) minium is given as the Latin *minium secundarium*. It is also given by Alphita as cinnabar, obviously incorrect.

[162] *bauraq*. From Pers. *bawra* or *būra*. An impure sodium carbonate, it also sometimes designates impure borate. The ancient Mesopotamians knew many types of impure soda which had various uses (Levey, 72, 105, 121, 128, 152). It was used as a dyeing auxiliary (Levey, 173) and for its detersive properties (Levey, 124, 173). See discussion on ancient natron (Levey, 121). Diosc. (V: 113) knew soda (νίτρον) in various varieties. In Arabic times (Maim., 51 and al-Ghāfiqī, 183), it was well known in ophthalmology (Ḥunain, 158, 163, 175, 178) and in ancient Egypt for washing and glassmaking. In *Aqrābādhīn* (fol. 124*a*) it is used with other simples in a clyster. The *Tuḥfat al-aḥbāb* (92) says its synonym is *nitrūn*. I. B. (381) states that al-Rāzī used it with oil of jasmine on the genital organs to strengthen them and provoke intense erection. Among other uses ibn Sīnā used it as an anthelmintic. Aristotle's lapidary (173) claims that *bauraq* removes the slime of the stomach in a compounded form. Natron (*ibid.*, 47) is given as a type of borax used for its impurities. There is thus some confusion regarding the properties of *bauraq*. Much of the bauraq used was probably a mixture of soda and borax.

[163] The yellow sulphide is designated by "arsenic" throughout the text.

[164] *khall khamr*.

[164a] A Samqarand village according to Yāqūt (*c.* 1179–1229) in *Muʿjam al-buldān* (Beirut, 1955).

[165] *ḥibr raiḥān*. *raīḥān* = *bādharūj*. Diosc. gives it as good for the eyes and as a carminative. I. B. (223, 892) states that al-Rāzī and ibn Sīnā used basil as a cardiac stimulant, and for the bile, among others. The *Tuḥfat al-aḥbāb* (72, 179) discusses *badaranjūya* and *ḥamāḥim* or *ḥabaq*. It is probably the latter which is identical with *raīḥān*, the grand basil, *Ocimum Basilicum* L. *Cf.* Diosc. (II: 141), ὤκιμον.

[166] *al-nīl*, sometimes called *al-nīlāj*. Probably *šamlalangu* in Akk., perhaps related to Pers. *līlanj* = indigo plant (*DAB*, 107; Levey, 105). The more common word in Persian is *nīlah*. The Indigofera has been known in India for thousands of years, particularly in medicine (Ainslie **1**: 178–180 for hepatitis and nephritis. Diosc. (V: 92) included it in his materia medica, mentioning its use for inflammation and oedemata of ulcers, and also its importance in the dyeing industry. The Arabs knew it well. I. B. (1562, 2244) mentions its use for palpitations and gangrene by the Muslims. In North Africa (*Tuḥfat al-aḥbāb*, 292), it has been and is used mainly as a dye and is called *nīla*. *Indigofera tinctoria* L. is still grown in Persia. It is called *rank-i-kirmānī* although the ancient name for the indigo leaves in the Punjab, Iran, and Turkey is *wasma*, the name formerly used for woad, a dye obtained from *Isatis tinctoria* L. These two plants were often confused in the ancient literature (Hooper, 129).

ammoniac[167] are soaked in water of the sapanwood[168] tree a day and a night. On the morning it is kneaded with the fingers in the vessel in which it is. It is then clarified. On it are thrown three dirhams of saffron to obtain a stronger color than gold. Its yellow is prettier.

Preparation of a blackish ink. One part of bee honey[169] is taken and one part of mica[170] and one part of *qalqand* vitriol. The vitriol, mica, and honey are pounded and put in a cucurbit and alembic[171] and distilled. Then the distillate is removed from it into a vessel and placed in the sun for twenty days. Every day there is pulverized for it a dirham of gum arabic which is put into it. It is stirred vigorously until the gum is dissolved. When one writes with it, it comes out beautifully.

Preparation of another. One part of vitriol and one part of green vitriol are pulverized together. With it is a bit of gum. It is dissolved in water of boiled gallnut. Use it.

Preparation of sumac ink. One half *raṭl* of sumac[172] is taken. On it are poured three *raṭls* of pure water. It is placed in the sun for two days until the red of the sumac comes out. It is squeezed, filtered through a fine cloth, and put in the sun for five days. On every *raṭl*, five ounces of gum are put—every day an ounce. It is left until the gum is dissolved. On it is thrown what is necessary of vitriol. It is checked so that it is not burned owing to an excess of vitriol. (**24**) Use it.

Preparation of an ink with which one writes so that it comes out white on black and black on white. It is strange and nice. Four dirhams are taken of the best soda.[173] On it is poured one half *raṭl* of water. It is stirred, then left for seven days. As often as the water diminishes, there is added to it an amount that disappeared from it. It is stirred. After the days pass, the water is clarified on eight dirhams of collyrium, i.e. collyrium of the pulverized dirhams and three dirhams of marcasite,[174] and a dirham of myrrh. If it is spotted, it is pulverized in a mortar one day and there are added to it four dirhams of vitriol and two dirhams

---

[167] *washaq* or *wasaq*, sometimes *ashaq*, *ashaj*, or *washaj*. It was known to Diosc. (III: 84). In India (Ainslie **1**: 158–160), known from the Arabs and Persians, gum ammoniac was used as a discutient and resolvent; internally, it found use as a deobstruent and expectorant. It is listed in the *Tuḥfat al-aḥbāb* (29, 135) as the "gum of *el-kelkh* and one calls it *el-fāsūkh*." *Cf.* Maim. (124). Ducros (174) lists it as *fasūkh*, a product of some species of Ferula.

[168] *baqqam*. *Caesalpinia Sappan* L., sapanwood, was well known in Arabic times. I. B. (314) quotes ibn Ḥassan to the effect that the powder of the root taken in a dose of five drachmas is fatal. Al Bīrunī wrote that the name in Persian for sapanwood was *dar barniyān* and in Khwarizmian, *banjank*. Much of it came from Sumatra. It was used for cicatrization of wounds, to dry ulcers, and as a styptic (al-Ghāfiqī, 123). Al-Kindī used sapanwood in his perfumery recipes (15, 17). The *Tuḥfat al-aḥbāb* (315) gives *ʿandam* as the North African equivalent of *baqqam*, sometimes called *dam-al-akhawain* because of the red color.

[169] *ʿasal naḥl*. In Babylonia (Levey, 86, 94, 104), honey was a valuable product of bees. It was used in medicines, bread, and to make alcoholic drinks (*Cf. DAB*, 284). Diosc. (II: 82) gives it as μέλι having many uses in therapeutics. In Arabic times al-Kindī (8*a*) used honey in the distillation of saffron and rose waters. Jābir also used honey in chemical experiments in his book on poisons (63*a*). In I. B. (1542), the Arabs are shown to have used it for teeth, facial tic, nausea, and gums. Honey in Arabic sometimes refers to a gum or resin. A mixture of vinegar and honey is given in *Tuḥfat al-aḥbāb* (400) as *sakanjubīn*.

[170] *ṭalq*. Here mica but frequently talc. The mica gave the ink, when dry, a shiny metallic appearance. Mica and talc were well known in ancient Mesopotamia (*CAS*, 81). Aristotle's lapidary (50) considers *ṭalq* as a kind of asbestos. In this sense, it is close to that mentioned in Diosc. (V: 138). The latter, however, is ἀμίαντος λίθος, probably asbestos, *amiantus lapis*. In Arabic times, *ṭalq* denoted many minerals. Jābir used *ṭalq* (*cf.* II: 54*b*) as did al-Rāzī (46). I. B. (1472) considers *ṭalq* as a useful simple in the Arabic materia medica for tumors, hemorrhages, the anus and dysentery. I. B. is confused as to what *ṭalq* really is. The *Tuḥfat al-aḥbāb* reflects this difficulty. The same confusion is in Steingass (818). Dozy gives *ṭalq abyaḍ* as mica, brilliant powder. *Cf.* Maim. (177).

[171] *qarʿah wa-anbīq*. Distillation apparatus. See al-Kindī (18 ff.) and al-Rāzī (54 ff.). For the earliest distillation apparatus, *cf.* Levey (chap. 4) and Levey, M. The earliest stages in the evolution of the still, *Isis* **51**: 31–34, 1960.

[172] *summāq*. According to a Babylonian text (Levey, 74), the oak and sumac were two of the principal sources for tanning agents. Sumac was used medicinally in Babylonia (*DAB*, 167, 239, 241). Diosc. (I: 108) considered sumac, ῥοῦς, *Rhus coriaria* L., as a tanning agent to dye hair black, for dysentery, and for pain of the gums and teeth. In Arabic times, I. B. (1217) explains sumac was used by al-Rāzī as an astringent, for uterine hemorrhage, and polyurea. Others used it in ophthalmology, to prevent inflammation, and for contusions. In the *Tuḥfat al-aḥbāb* (368), the Berber term for sumac is given as *tizgha* (sing. *tazeght*). Ducros gives the synonyms of sumac as *summāq al-dabbāghin*, *summaqīl*, and *tamtam*. It is sold in the Egyptian bazaars as an astringent, antihemorrhagic, and antiseptic. Sumac still grows in the Near East. It is still used to dye silk and tan leather goods by the Arabs, Turks, Iranians, and in Europe (Hooper, 164; Nabat, 200; ibn Rasul, 566).

[173] *qalī*. Well known in Babylonia and to Diosc. (V: 119). *Cf.* note 162. This is the most common Arabic word for soda which is an ash from alkaline plants. The word "alkali" is derived from *qalī*. Al-Rāzī, *passim*, discusses soda and its many uses. See also Alaune on *salis alkali* (52ff, 81ff.) and Maim. (345). *Salicornia fruticosa* L. was the plant most used to obtain *qalī* in Babylonian as well as in later Arabic times (Levey, 121). For the preparation of alkali from plants in India, see Ray (63–64). For *qalī* in Jābir, see his book on poisons (II: 54*b*; IV: 105*a*, 112*b*).

[174] *marqashīta*. Marcasite, to ancient writers, is variously a mixture of iron pyrites, bismuth, and antimony. Later it referred to FeS_2. It was used in the ancient Near East for ornaments (*CAS*, 117). Aristotle's lapidary (24) states that "there are many types—gold, silver, and copper marcasites." (*Cf.* Wiedemann **43**: 97). It is useful in chemistry when it is calcined and burned until it is a fine powder. When a bit of it is brought together with sulphur in a crucible, it purifies gold (Aristotle 24). Al-Kindī knew marcasite and discussed it in his work (II: 55*a*). I. B. (2116) states that al-Rāzī used marcasite for the eyes and in a compounded recipe for leprosy. *Tuḥfat al-aḥbāb* (262) calls it "stone of light because of its uses for the sight. Its types are the gold, silver, copper, and iron." It is noted that ibn Sīnā and al-Razzāq state the same thing. Ducros (217) notes the sale in present day bazaars in Egypt of *marqashītā dhahabyah*, a pyrite, for kidney calculi.

of white lead.[175] It is all left well pulverized. On it are poured three ounces of water. It is left for five days, then it is boiled. Then there is taken of water of soda, and of the collyrium—one ounce. Then water of soda and collyrium are boiled, ounce after ounce, and also five dirhams of crushed gallnut. It is boiled until one-third has disappeared; two-thirds remains. It is then clarified and mixed. If there are iron filings with the gallnut, then it is good. Then the water is mixed with the previously mentioned things. The two waters are mixed together and with it a bit of gum arabic and starch paste. Write with it on black; it comes out white. On white, it comes out black.

Preparation of an ink with which one writes like gold. Six *mithqals* of lead are taken. (25) On it are thrown four *mithqals* of vitriol. Then it is placed in a luted vessel and put in the higher part of a glass furnace a day and a night. It is then taken out. On it is poured water of gallnut. Write with it. When polished, it comes out a very good golden color.

Preparation of another ink that is golden like it. A male goat's gall is taken. Write with it on new paper with a good pen. It comes out as gold.

Preparation of a rose colored ink. Two parts of white lead are taken and one part of red lead.[177] They are put together with vinegar in a new pot which is then luted with clay and hair. The pot is put into a glass furnace in the upper part for three days. It is then taken out and the material pulverized. On it is poured water of white gallnut. A bit of gum arabic is put into it. Write with it.

Preparation of a monk's ink. The strong red leaf of the anemone, from which the black part has been removed, is then boiled in water until its color comes out into the water as desired. Then it is taken down and purified. To it is added water of myrtle, in an amount of one quarter of the water, and two dirhams of gum arabic. Write with it.

Preparation of another green ink. White gallnut is taken and crushed lightly. On it is poured enough water to cover it. It is left for an hour or less to the extent that it takes the strength of the gallnut. It is then purified. Then the best pure green verdigris is taken in the amount desired and pulverized well. On it is poured a little wine vinegar. It is kneaded and put on a brick until its moistness disappears. Then it is well pulverized. The pulverization is the essence of the work. On it is poured water of gallnut. (26) It is well beaten and left. In it is put powdered gum arabic in the amount desired. It is then stirred. Write with it.

Preparation of a yellow ink. Water of gallnut is taken in an amount like that taken for the green, and instead of the verdigris, yellow arsenic is put in but with no vinegar in it. It is then beaten with water of the gallnut and a bit of starch glue. If water of bran[178] is put in, then it is better.

Preparation of a white ink. Gallnut is taken and crushed lightly. On it is poured enough water to cover it. It is left for one hour with a quantity slightly more than it. Then sieved powdery white starch[179] is pulverized well with water until it becomes as one phase in appearance. It is then left until it is clarified. The upper layer is taken and the residue is left. Then gum arabic is taken and pulverized in the water that was taken previously and in the starch glue. When it is dissolved, then it is beaten with that residue obtained. It is stirred and then put aside. If it is desired to work with it, then it is stirred. Write with it.

Preparation of a pretty red ink. The same amount of water of gallnut is taken like that used for white ink. It is put aside. Then red cinnabar is taken and washed.

Description of washing[180] of the cinnabar. It is washed when water is poured on it while it is in a vessel and it is being stirred. When a foam is raised, it is assumed that the impurities do not remain in the solution. Then it is put on a brick until the dampness is gone. It is then powdered until it becomes like paste. It is then beaten with water of gallnut that was put aside. It is left an hour. The gum arabic is taken, dis-

[175] *isfīdāj al-raṣāṣ*. Aristotle (70) states that "it is made from lead and vinegar and is healthy for the white in the eye." It is mentioned in Babylonian texts (*CAS*, 9, 106) and in Diosc. (V: 88*a*). Jābir mentions white lead in his book on poisons (II: 54*a*; IV: 111a; VI: 185*a*). *Isfīdāj* is a Persian word and is probably equivalent to the ψιμύθιον of Diosc. (V: 88). *Cf.* also al-Ghāfiqī (109). It is not listed in Ducros. In India, it has been known for a long time (Ainslie **1**: 534–535).

[176] *atūn al-zajāj*. A furnace for glassmaking, already known in the Babylonian period.

[177] *usrunj*. It was well known in Babylonia (*CAS*, 7, 8, 10). In Aristotle (181), "This stone comes from lead (*raṣāṣ usrub*); it is changed to become red when the fire has an effect on it. When made into a salve, it heals wounds. . ." It probably was known in many types, each having its peculiar impurities.

[178] *nukhālah*. *Cf.* al-Bustān. Bran extract was used in dyeing in Babylonia (Levey, 109) in the first millennium B.C. Before dyeing, wool was first steeped in bran extract (or its water). Then it was warmed gently with the dye and mordant (*Talmud Babli, Tos. Shebuot*, 5, 8, 68). See also *Tuḥfat al-aḥbāb* (107), *jummār* as a synonym for *nukhālah*. I. B. (2219) gives bran as *nukhālah*. It is to be found in Diosc. (II: 85), πυροί, applied as a cataplasm for leprosy and inflammations.

[179] *nashāstaj* or *nishāstaj*. From the Persian *nishāstah* (Steingass, Dozy). Also called *al-nashā'* (*Tuḥfat al-aḥbāb*). It is the starch from wheat. Diosc. (II: 101) cites it, ἄμυλον, as does I. B. (2224). The Arabs used starch as an anodyne, styptic, and astringent (Ainslie **1**: 404). Al-Kindī used *nishāstaj* in a number of his recipes in perfumery (Nos. 17, 18, 23, 25). *Cf. Tuḥfat al-aḥbāb* (289). Al-Rāzī thought it engendered obstructions but it was useful for the chest and lungs and good to combat flow in a cold in the head. Maim. (261) also listed it in his materia medica with the synonyms *al-nashā'* and *amtūlūn*.

[180] The cinnabar is washed by adding water, then stirring well. The Arabs assumed that it was the froth which always held the impurities. This fact is related to the early development of metallurgy. The froth together with the impurities and the water were then decanted. This was a very common operation. The washing operation is described well in Alaune (40) for arsenic. Salt was sometimes added to give the water a taste so that the chemist would know when all the water had been washed from the arsenic. This involved repeated washing with "sweet water."

solved in water, and thrown on it, then beaten vigorously (27) Write with it.

FIFTH CHAPTER
ON THE PREPARATION OF *LĪQ*

Description of red *līq*. There is taken of distilled soda[181] as much as is desired. To it is added washed, powdered red cinnabar which one deems sufficient in weight according to the eye. It is then put into a clean vessel and on it is poured enough water of fresh unused sapanwood to cover it. The *līq* is made from it. Write with it.

Preparation of a beautiful red *līq*. One part of red lead is taken and one part of Indian indigo. Each is powdered separately. Then they are put into a clean vessel and water of gum is poured over them. Write with it.

Preparation of a *khalūq*[182] *līq*. One part of red lead is taken and one part of yellow arsenic. Each is powdered separately. Then one is added to the other with strong pulverization. Water of gum is added. Write with it.

Preparation of pomegranate flower[183] colored ink. What is desired of green gallnut is taken. It is crushed with an equal amount of strong vinegar. Then it is left to settle to become well clarified. It is mixed with a bit of saffron and then boiled with powdered gum arabic. Then it is used.

Preparation of pistachio[184] *līq*. Ten dirhams are taken of celandine[185] (probably madder here) and water poured on it to immerse it. This is done in a small heating pot. It is cooked until when feathers are immersed in it they will be dyed. It is taken down from the fire. The water is decanted from it. (28) Then a dirham of hairy saffron, complete with its hair, is put in the clean water. It is boiled until feathers can be dyed. When the goal is achieved, it is then well purified. Water of rush[186] and water of rinds of the pomegranate are taken in amounts than can be taken up and put in. It must be the proper amount. Then two dirhams of pulverized, sieved gum are added. Write with it.

Description of a beautiful green *līq*. Gallnut is crushed and water poured over it, enough to cover it. It is left an hour until a little of the strength of the gallnut goes into the water. It is then purified and set aside. Then the best red cinnabar in an amount necessary is washed in water. This is done by pulverizing it, pouring much water on it, shaking it, and then transferring it to another vessel until its foam has come out. It is left to settle. It is purified until there does not remain in it a bit of water. It is then set aside until it is dry, i.e. its dampness disappears. It is then pulverized until it becomes like paste. It is beaten with some of the water of gallnut which had been set aside and in which two dirhams of gum arabic or whatever is necessary has been dissolved. It is all mixed. Write with it.

Preparation of a strong yellow *līq*. Yellow arsenic platelets and saffron, of each one part, are pulverized separately. They are then mixed by pulverization with the same quantity of gum arabic and placed in a clean vessel. Water of gum is then poured on it to cover it. Write with it.

Preparation of a beautiful blue *līq*. Two dirhams of celandine are taken. It is a twig found at the dyers. It is put in a pot and cooked as has been previously

---

[181] *usnān* or *ushnān*. This is an impure alkali from plants. Ritter, H., J. Ruska, F. Sarre, and R. Winderlich, Orientalische Steinbücher u. persische Fayencetechnik, *Istanbuler Mitteilungen* **3**: 33, note 4. See earlier note on soda. In Berber, this is called *tāsrā* (*Tuḥfat al-aḥbāb*, 38) and in Persian *ishnān* or *ushnān*. I. B. (87) cited *ushnān* as having many medical uses. Fuller's soda is given as *ghassūl* while *ushnān* = *ḥorḍ*. Many types of soda were available in the ancient and mediaeval periods. Owing to incomplete descriptions, it is impossible to know the impurities of the various kinds and their many sources. *Aqrābādhīn* (115*a*) mentions it for a clyster. Maim. (24) gave the Greek name as ἀδάρκης and Berber as *taghaïghaït*.

[182] *khalūq*. A mixture of perfumes including saffron as an ingredient (*cf*. Lane). It is mentioned in many works. Al-Kindī (42*a*) gives the ersatz formulas in nos. 24, 25. Saffron is usually added to spices pulverized in jasmine or sesame oil. Honey and drugs are then added. The Persian is *khulūq* (Steingass).

[183] *jullinār*. Persian = *-jullanār* and *julnār* (from *jul* "a flower" and *nār* "a pomegranate"). Gr. = βαλαύστιον. In Ar. it is also called *al-raghath* and *al-maẓẓ* (Maim., 75). In India the flower was used as a vermifuge and stomachic (*DAB*, 315). Diosc. (I: 97, 111) considered pomegranate flowers, κύτινος, to be good for agglutinating of wounds, for the gums, and for loose teeth. Whoever swallows the flowers will not be troubled with eye disorders for a year, some say. There are various species of this tree. The Persians consider the blossoms among the cicatrizantia (Ainslie **1**: 322–324. *Jullinār* comes from the male pomegranate which does not bear fruit and is sometimes called the "bad pomegranate (al-Razzāq, 205). Al-Kindī (35) used *jullinār* in a recipe to falsify saffron (nos. 17, 18). In the *Aqrābādhīn* (100*b*), it is used with other simples for sores. Zahrāwī (195) mentions rose tablets for stomach bleeding; for rose oil *cf*. Zahrāwī (209ff). I. B. (494) claimed that *jullinār* is good for an excess of humours in the stomach and intestines. The *Tuḥfat al-aḥbāb* (94, 287) gives a synonym as *rummān al-murūj*. Today, it is sold in the Egyptian bazaars (Ducros, 65) also under the name of *nārmishk* employed mainly as an astringent for external and internal use. *Cf*. al-Ghāfiqī (194).

[184] *fustuq* or *fustaq*. The Hebrew *boṭnīm* is cognate to *buṭnu* of Akkadian, probably pistachio. The Arabic word is probably related to the Pers. *pistah* (Lane). Diosc. (I: 71) gives τέρμινθος which is *Pistacia Terebinthus* L. as being good for what the lentisk is good for—for leprosy, pruritis, in cataplasms, and for other ailments. Al-Kindī (28) in recipe 47 used pistachio to prepare *ghāliya*. I. B. (1681) gives *fustuq* as "pistachier" while Sontheimer (II: 255) calls it *Pistacia vera*. The *Tuḥfat al-aḥbāb* (321) says that *fustaq* was imported from the Orient. It was, however, widely known in Syria according to I. B.

[185] *ʿurūq ṣabbāghīn*. I. B. (1525) says that these are yellow roots of the celandine to be found in Spain, Greece, and in the Berber countries. Dioscorides knew this plant (II: 180), χελιδόνιον, as good for the vision and the teeth. The greater celandine is the *Cheledonium majus* L. while the χελ. τὸ μικρόν, lesser celandine, is the *Ranunculus Ficaria* L. These two come under *ʿurūq ṣabbaghīn* in I. B. who knew both. I. B. gives an important synonym as *māmīrān*.

[186] *asal*. *Cf*. Lane. Diosc. (IV: 52) calls it σχοῖνος. The *Tuḥfat al-aḥbāb* (22) states that it is *al-sumār* (*cf*. Dozy). I. B. (65) cites its use in Galen for the head.

noted until the feather is dyed. (**29**) It is then removed from the fire and purified. Then water of indigo is added in an amount sufficient for it and for the color desired. It is then beaten with water of gallnut. Powdered gum arabic is beaten into it. Then carry on in what you wish.

Preparation of a yellow apricot[187] colored *līq*. Three parts of yellow arsenic and one part of saffron are taken. These are pulverized together and moistened with lukewarm water with gum and saffron until all is dissolved. It is also mixed with the yellow of an egg. It is put into white wool[188] *līq*. Write with it.

Preparation of green *līq* like emerald.[189] Verdigris is pulverized with an equal amount of white gum arabic in water of gallnut. Then a little wine is poured on it. It is then used.

Preparation of a green *līq*. Three parts of verdigris and two parts of gum are taken. These are pulverized together with wine vinegar very well and also one dirham of vinegar. Then write with it.[190]

Preparation of an apricot-colored *līq*. Yellow arsenic is taken as needed and pulverized with water of gallnut and water of gum until it is well done. It is then dried. Then one-sixth of its weight is taken of Iraqi indigo. These are pulverized with water of the leek[191] or with water of rocket[192] or coriander.[193] Use it.

Description of a white marble *līq*. White lead is taken as desired and pulverized with water of gallnut which has been soaked an hour. It is pulverized well, then dried, and water of gallnut is gradually introduced, judged according to the appearance. Then write with it.

Preparation of a *lāzward* colored ink. Balkh[194] *lāzward* in a quantity desired is taken and water is poured on it to cover it. (**30**) It is then stirred well and left overnight until it is clarified. Then the white water is decanted from it. Then water of soaked gallnut is poured on it together with gum. Write with it.[195]

Preparation of a golden yellow ink. Two parts of honey are taken, one part of mica, and one part of very good Cypriote vitriol. They are powdered together with the honey and put in a pot and ambix. It is placed on the fire until distilled. The distillate is taken, put in a vessel, and placed in the sun for twenty days, every day of which a dirham of gum arabic and mica is pulverized and added to it. It is vigorously stirred until the gum arabic is dissolved. It is removed after that. One can write with it as desired. It comes out the color of gold.

Preparation of another golden ink. One part of yellow vitriol is taken and also its fourth of ammonium chloride.[196] The vitriol is pounded but it is left coarse. The sal ammoniac is pulverized with it to mix them. It is bound in the bladder of an ox with the head tied. It is

[187] Yellow arsenic may be used without saffron in a *līq* preparation, or "if the yellow arsenic is not available, Iraqi *isfīdāj* is used with saffron and gum" (MS Berlin 5567, fol. 53*a*).

[188] *ṣuf*.

[189] *zabarjad*. The type of emerald meant here is highly uncertain. Lane gives chrysalite but this is modern, not ancient. Aristotle's lapidary (2) describes the emerald, *zabarjad* or *zumurrud*, in use as a poison. "It is a pure green stone; the noblest are those which are dark green." Emerald is also mentioned in Alaune (117). *Zumurrud* is given in I. B. (1123) as is *zabarjad* for emerald. It is useful in leprosy according to I. B. The *Tuḥfat al-aḥbāb* (159, 162) gives it as topaz or emerald.

[190] Another green *līq* is made from yellow arsenic and Indian indigo with gum added. It is more difficult to obtain the desired color from indigo (MS Berlin 5567, fol. 53*a*).

[191] *kurrāt* or *kurrāth*. The meaning of this term is still uncertain. It may be leek which in Sumerian is *GA.RA Š. ŠA R* equated in a list to the Akk. *karashu*. This probably is *Allium porrum* L. The latter is given in Diosc. (II: 149), πράσον, for the belly but causes troublesome dreams. It causes dullness of sight, hurts the ulcerated bladder and kidneys. It is, however, used for other purposes in medicine. *Aqrābādhīn* (124*a*) gives it for hemorrhoids when used with walnut oil. I. B. (1910) discusses the Syrian leek. *Cf.* also I. B. (1911). Maim. (198) has the synonyms as *balābis*, and for the Syrian leek *al-kurrāth al-shāmī*, and for the cooking leek, *al-kurrāth al-bustānī*, also called *al-qaflūt*. *Cf.* also *Tuḥfat al-aḥbāb* (83).

[192] *jirjīr*. It is also *al-kathā* in Arabic (Maim. 74). It was known in Babylonia as *gīrgirū*. This is cognate to Syriac *gargīrā* and to Ar. *jirjīr*. Usually, it is the *Brassica eruca* L., used for the eyes and as an aphrodisiac (*DAB*, 212). Diosc. (II: 140), εὔζωμον also describes the rocket. I. B. (473) describes the drug from the rocket of Galen, Dioscorides and others. *Tuḥfat al-aḥbāb* says that it is called *buk ᶜAlī* "your father ᶜAlī."

[193] *kuzbara*. Coriander was well known in Babylonia. In Akkadian, it is *kisibirru*, cognate to the Arabic and to the Aramaic *kusbartha*. It is probably *Coriandrum sativium* L. (*Cf. DAB*, 66; Levey). It was known to Diosc. (III: 63), κόριον, to heal ulcers, carbuncles, and inflammations of the skin. Later, it was known to Maim. (358) as "coriander of the fox" (*kazburat al-thaᶜlab*), and "wild coriander" (*al-kazbur al-barrī*). Maim. probably did not mean the coriander first mentioned. I. B. (1926) devoted much space to *kuzbara*, thus indicating that it was well known to the Arabs and Persians. The *Tuḥfat al-aḥbāb* (230) says that it is *al-quzbūr* with a *qāf* in the popular language. Ducros (199) lists it as still being sold in the bazaars as a carminative, digestive, stomachic, and condiment. *Cf.* note 160.

[194] Balkh was one of the capitols of Khurāsān (Transoxania), about ten parasangs or leagues from the Oxus or Jīhūn (Steingass).

[195] For a *lāzward līq*, for every dirham of white lead, use one-fourth of Indian indigo (MS Berlin 5567, fol. 53*a*).

[196] *nushādir* or *nūshādir*. Sal ammoniac was known in ancient Mesopotamia. It was called (probably) in Sum. *IM.KAL.LA* (*DACG*, 24 ff.) since it is obtained as a sublimate from the soot (*IM.KAL* = sublimate) of camel's dung. This was still true in the nineteenth century in Algeria. In Akk., the Sumerian *IM.KAL* is equated to *isikku* in a lexical list. This word seems to be related to the Syriac *assāka* = *sublimatio* from *assek* = to sublime. It was probably also obtained by fractional crystallization of the crystals found in waste drains just as it was in India. Sal ammoniac (*chulikalavana*) with sulphur and other materials was digested with cow's urine to make a *vida* for killing gold and other metals (Ray, 157). Rasaratnasamuchchaya states that *navasara* (sal ammoniac) is "produced by decomposition of shoots of bamboos" and of the wood of *Carya arborea* (Ray, 176). It "kills mercury," "liquifies iron," is a stomachic, an absorbent of the spleen, and aids digestion after much eating. Aristotle's lapidary (45) mentions *nūshādir* to make dyes fast. Sal ammoniac was known to Jābir (IV: 113*a*, *b*). Sal ammoniac is mentioned frequently in al-Rāzī (*passim*). Al-Kindī used *nūshādir* in his recipe no. 67. I. B. (2241) discusses two species, a natural and an artifical salt. It is good for the bath and for leucorrhea. *Cf.* Ruska, J., Sal ammoniacus, Nusadir und Salmiak, *Sitzungsb.d. Heidelberger Akad. d. Wissen., Phil.-hist. Klasse* Part 5, 1923.

hung in an oven having a low fire for one night. It is covered. When morning comes, it is removed. It is then found that what is in it has become creamy, thick, and is viscous. Write with it on cloth, parchment, or what you wish.

Preparation of a silver *līq*. A *raṭl* of the best mica is pulverized and put in a vessel which has been untouched by fat. In it are put ten dirhams of *tutia*[197] and on it sharp pure vinegar is poured, enough to cover it by a finger.[198] It is placed in the warm sun for fifteen days. It is removed from the sun and put in a thick Kordovan cloth bag. Warm water of cooked beans[199] is used in which to press the bag. In it has been put a little saffron. It is rubbed vigorously with the palm of the hand. To what comes out from it, pulverized saffron and pulverized gum arabic are added. **(31)** Then write with it for the color of gold comes out. If a silver color is wanted, then it is used without the saffron but with the gum. It comes out silvery.

Preparation of *khalūq līq*. The desired amount of mica is cut up with shears until it is smaller than the grain of a mustard[200] seed. It is bound in a thick cloth and rubbed until whatever is necessary comes out. It is sieved with another thick cloth. Then one part of it is taken and one part of red arsenic which has been well pulverized and one part of the mentioned mica. It is all gathered and kneaded with water of gum arabic which is described later. It is then dried to the extent desired. It is removed. If it is desired to write with it, a small amount is taken or whatever desired and dissolved in a shell container with water of gum. Write with it. However, if a golden color is desired, in place of red arsenic, yellow is used. The color is then yellow.

Preparation of water of gum with which the colors are mixed and also others. A *raṭl* of clarified gum arabic is broken into pieces. On it is poured clear water. It is then boiled on a low fire until it is dissolved and has the consistency of honey. It is put in enough water to cover it. When it is cooled a little, it is used.

Preparation of a golden *līq* from anemones. The black part of the petals of anemones is cut off and thrown away. The red is removed and collected in a pot. On it is poured enough water to cover it. It is put on the fire and boiled until its color comes out into the water according to what is desired. It is then taken down and purified. Two dirhams of water of myrtle are thrown on it and also a quantity of gum arabic—**(32)** as much as one quarter of the water. Write with it.

Preparation of a rose *līq*. One part of white lead and one part of red lead are pulverized in wine vinegar and then put into a little pot luted with clay of the art. It is placed in the upper part of a glass oven and left for three days. It is then removed, powdered, and on it a little water of gallnut is poured. A bit of gum is added. Write with it.

Preparation of violet[201] *līq*. Ten dirhams of celandine are taken. On it is poured enough water to cover it in a small pan. It is cooked until it disintegrates. It is taken down and its water purified. Ten dirhams of hairy saffron are put in the water. It is now in the proper portion. It is then boiled until the feather is dyed and then well purified. Then water of myrtle or water of the rind of the pomegranate in the proper amount is added; if it is more, then it turns black. Let it be in a pot. Then two dirhams of sieved gum arabic are thrown into it. Write with it.

Preparation of another *līq*. One part of yellow

---

[197] *tūtiyā*. According to Aristotle's lapidary (52), this stone is found in mountains. "There are many types, white, yellow, and green. All are useful for moisture in the eyes." *Tutty* is mentioned in Diosc. (V: 75) as being used to cicatrize malignant and creeping sores and to dry abscesses. There is a great deal of uncertainty as to what *tutty* is. For the washing of *tutty*, *cf.* Zahrāwī (173). Maim. (382) gives the synonyms as *iqlīmiyā al-ṣufr*, *qashqaṭūta*, and *qadmiyā*; thus it would appear to be *cadmia* in a very impure state. Its color would depend on its impurities. I. B. (437) says that *pompholyx* is *tutty*. According to the *Tuḥfat al-aḥbāb* (403), the best is the white type. The word *tūtiyā* perhaps comes from *tūt* "mulberry," thus coming from the Greek word. *Cf.* also al-Rāzī *passim* for tūtiyā.

[198] *asba*ᶜ. A measure of length.

[199] *bāqilā*. The Akkadian synonym is not yet known. In Diosc. (II: 105), κύαμος Ἑλληνικός, it is the *Vicia Faba* L. called "Greek bean." It had many medicinal uses—for eye difficulties, and in cataplasms for children. It also dyes wool. On the other hand, it causes "ugly and false dreams." "It also increases the flesh of the body, and when cooked in vinegar and water and eaten with its peel, it checks diarrhoea caused by ulceration in the gut" (al-Ghāfiqī, 127). Maim. (47) stated that it is the *al-jirjīr* with the more common name being *al-fūl*. The former name is doubtful. The Aramaic *fūlā* probably is the origin of *fūl*. The *Tuḥfat al-aḥbāb* (76) also gives *al-fūl* as the synonym. The *fūl* is today grown in Iran and in other parts of the Near East (Hooper, 184). The Hindu for it is *bakla*, and in Turkish *baqlah*, in Kurdish *paglah*, and in Persian *bāqilā'*. I. B. (224) cites the uses of *bāqilā* in Galen, ibn Sīnā, and others.

[200] *khardal*. Mustard was a very common condiment and simple in ancient Mesopotamia. *Cf.* Levey (230) for its use in baking, beer, and in seasoning; *cf.* also *DAB* (203 ff.) for the medicinal uses. The Sumerian for mustard is ^{šam}*ḪAR.ḪAR*, in Akk. *khaldappānu*. It was used particularly in venereal disease, hemorrhoids, for the eyes, in enemas, and as a sialogogue. Some of the synonyms for *khaldappānu* are *khasisanu*, *khallamesu*, *khalulaia*, and *sappandu*, and in Sumerian *TUR.RA* and *RU.UŠ.RU.UŠ*. Thompson (*DAB*, 207) proposes that Ar. *khardal* may have come from Akk. *khaldappānu* (consider the Syriac *ḫardhᵉlūna*). The Mishnaic Heb. is *khardel*. In Diosc. (II: 154), σίνηπι is good for the head, swellings, and leprosies. In *Aqrābādhīn* (121*a*, 123*b*), it is used for burns and leucorrhea. Maim. (400) states that the white type is *isfandār* and the wild *al-ḥarshā'*. The name of the entire plant is *al-ḥarshā'* and that of the seed *al-khardal*. The *Tuḥfat al-aḥbāb* (417) calls it *bū ḥammū* in Berber. The mustard (*Brassica nigra* Koch) and other species are today abundant in Iraq and Iran.

[201] *banafsaj*. Diosc. (IV: 121) gives it, ἴον, as a stomachic and a help for convulsions in children. Carnoy shows that the Greek may have come from the Indo-Eur. *u̯ei*, "to turn." Jābir knew the violet as did al-Kindī who mentions the violet oil in nos. 35, 58, 81. He obtained the oil by distillation. *Aqrābādhīn* (103b) employs it in a plaster for the spleen or stomach. It is probably *Viola odorata* L. I. B. (353) says that it is a well-known plant employed medically. The *Tuḥfat al-aḥbāb* (note to 63) gives the Persian synonym as *janofshah*. Meyerhof states that Maimonides excluded *banafsaj* from his list (*cf.* Maim., 62) since it was too common and well known.

vitriol, five parts of pulverized Cypriote vitriol, and water of gallnut are pulverized with water of the gallnut and then put in a glass vessel. Its mouth is luted. It is then buried in dung for four days. Then after that, its contents are dissolved in water of gum and water of ammonia. Write with it.

Preparation of a white pretty *līq*. Two parts of white lead, its equal of mica, two and a half dirhams of gum, and the same amount of gum tragacanth[202] are taken. It is all powderized. Fish glue[203] is put into it. Write with it.

Preparation of a black *līq*. (**33**) Three parts of fresh walnuts,[204] before they are formed, are taken and one part of vitriol. They are pounded with some gum arabic and dissolved in water of the gallnut mixture. Use it.

Preparation of gold *līq*. One part of blue vitriol is taken, one part of mica, three parts of honey and put in a vessel. Its head is luted with clay. It is buried in a dung fire for seven days. It is then taken out and put in an alembic. It is distilled with gum arabic in it. One writes with it.

Preparation of another excellent *līq*. Gold is filed, then placed in a clean container. On it is poured enough vinegar to cover it. When it is dissolved, the vinegar over it is filtered off little by little. Fish glue is added to the residue. Write with it. The pen is dipped in alum water.

Another *līq*. White lead is taken and made stringy. Empty it into sweet water. Melt the obryzum[205] and empty it into the mixture. Then it is found softened. It is then pulverized on a stone and mixed with water of gum. Write with it.

Preparation of another red color. A dirham of a very good red earth, called a "red vein," a *dāniq*[206] of gum arabic, and a *dāniq* of resin of tragacanth are pulverized together. The cooked pure water of lac is added. Work with it as desired. If it is desired to have it shaken, the solution is agitated with the hand.

Preparation of the odoriferous verdigris *līq*. Good old verdigris is pulverized on a stone with good vinegar free of oil. It is well pulverized. In it is put powdered gum in the measure needed. (**34**) It is removed in a clean *līq* in a glass vessel. When it is dry, it is necessary to moisten it. This is done with vinegar. Water is never used or it would spoil.

Preparation of a blue-colored *lāzward līq*. Old *lāzward* is pulverized with water on a stone. It is then collected in a glazed or glass vessel. Sweet water is poured on it. It is then left one or two hours until the *lāzward* remains in the lower part of the vessel. The water is decanted from it. Sweet water in the quantity of a full pitcher is added. It is shaken and left an hour until it is settled. Then that water is decanted from it. This is done three times until very little water remains. The gum is made as has been previously described, or use cooked fish glue.

Preparation of green *līq*. Yellow golden arsenic is pulverized well with water on a stone. Then very good indigo is thrown on the arsenic and well pulverized with it. It is then put in a *līq*. Write with it.

SIXTH CHAPTER[207]

ON THE MIXTURE OF DYES, COLORS, AND THEIR PREPARATION

Know that the colors are white, black, red, green, yellow, and the color of the sky.[208] The white is the *bauraq*. The black is the soot ink. *Lāzward* is the color of the sky with indigo, and verdigris is green. Red is made with cinnabar and red lead. The shining yellow is from yellow arsenic, and the red is from red arsenic. As to the dyes, some of them should not be mixed with some others unless they have been pulverized and moistened. It is better when it is so. Through the white lead, which is *bauraq*, a multiplicity of dyes is obtained. It changes from tint to tint. It alone is used for a good white and not anything else. The arsenic and the *lāzward* do not mix with anything else. There is not in them anything outside of their color. (**35**) There is a *lāzward najūbi*, made by taking one part of lazward and one part of *bauraq*. They are pulverized together.

---

[202] *kathira*. Tragacanth gum in Babylonia (*DAB*, 272) since it has a flaky appearance, was associated with scales of copper (*arkhu*) and scales of iron (*iarakhu*). The plant was *šamiarkhu*. Diosc. (III: 20) knew it as did the Indians (Ainslie **1**: 162–163). It was supposed to have a good effect on the state of the blood. Maim. (191) states it is the gum of the herb *al-qatād*, also called *al-sahāj*. I. B. (1889) states that it is abundant in the mountains of Lebanon. The gum was employed on ulcers of the eye and skin, and against purulent ophthalmia. It is also a purgative. The gum comes mainly from *Astragalus gummifer* Lab.

[203] *ghirā' samk*.

[204] *jawz*. This term may indicate many types of nuts. In general, however, when it is not modified, it is the walnut since this was so common in ancient times. Diosc. (I: 125) mentions the walnut, καρύα βασιλικά, *Juglans regia* L. as an antidote for poison when used with figs and rue. It is also used for suppurations and alopecia. *Jawz*, originally a Persian word, was well known in India (Ainslie **1**: 463–465). The Greek √καρ, tough {fruit}, probably comes from the Sansk. *karaka* (Carnoy, 66). The *Tuḥfat al-aḥbāb* (98 ff) and Maim. (82) list a number of nuts important to the Arabs. I. B. (525) quotes al-Rāzī who stated that the walnut provokes pustules in the mouth and inflammation of the tonsils. Al-Kindī (32) used walnut oil to extract oil of cotton seed and apricot seed.

[205] *ibriz*. Pure gold. On use of gold in chemistry, *cf*. al-Rāzī (246 ff.), also Alaune (97 ff.).

[206] *dāniq*. A unit of weight.

[207] For some Latin sources, *cf*. Thompson, D. V., Jr., Medieval color making: *Tractatus qualiter quilibet artificialis color fieri possit*, *Isis* **22**: 456–468, 1935; same author, Trial index to some unpublished sources for the history of mediaeval craftsmanship, *Speculum* **10**: 410–431, 1935. Only one of these sources goes back to the twelfth century, two each are in the thirteenth and fourteenth; the rest are dated later. The manuscripts listed contain descriptions of materials used in painting and writing, methods of polishing, gilding, gluing, sizing, and manufacture of various kinds of ink, writing instruments, parchment, and leather.

[208] "There are four basic colors for compounding the yellow of yellow arsenic, the red of cinnabar, the black of Indian indigo, and the white of white lead." (MS Yale L-379, fols. 7*a*, *b*.)

Then a small part of *bauraq* is introduced, little by little. It is changed from tint to tint. Take from it what is wanted.

Another color which is dark. One part of dry good indigo is taken and one part of *bauraq*, mixed, and powdered well together. Then one part of white lead is added. It changes every time a little is added until it comes to the desired color.

Section on the cinnabar colors. The color of cinnabar is the color of enriched turquoise.[209] A desired amount of cinnabar is pulverized well with grape vinegar until it can no longer be felt. Nothing else is mixed with it.

Another color inferior to it. Two parts of cinnabar and of *bauraq* are taken together and pulverized. *Bauraq* is added little by little until the color called *qirsh*[210] is obtained. It is whitish.

From it the enriched pottery glaze is obtained. Three parts of cinnabar are taken and one part of *lāzward*. These are pulverized together. Use it.

Section on the green color. Ten parts of yellow arsenic are taken with two parts of good indigo. They are well pulverized together so that it becomes a saturated **(36)** rich green. Every time it is desired to enrich the color, particles of arsenic must be added, little by little, until the noble green is achieved. There are many tints of this.

A color like the color of blood. The best red cinnabar is taken and pulverized in water. It is then left under cover until it is still. The white which rises is decanted. Water is then added to it and then decanted until the water remains clear. The residue is the color of blood. The cinnabar may be pulverized in water and salt. In that case, the upper layer is black. When settled, the black water is decanted from it. More water is returned to it, with pulverization, settling, and decantation. This is done until the water is pure. The cinnabar is tasted. If a salty taste is not found in the layer, then it has been attained. Use it.

Then there is a rose color. Three parts of *bauraq* are taken and one part of cinnabar. These are mixed together by pulverization. Each time one part of *bauraq* is added, it whitens further until it is returned to its original.[211]

Another color is orange.[212] The best red lead is pulverized well with water when needed. Write.

Another color is red ruby[213] from the *lukk*. How it is made. Description of how to dissolve the *lukk*. Ten ounces of *lukk* are broken up after it has been freed of its twigs. Then two dirhams of *ushnān* and two dirhams of *bauraq* are pounded very finely. Enough water is poured on to cover them. It is brought to the fire with the *lukk* until all of the redness of the *lukk* is brought out. **(37)** It is removed from the fire. It is filtered, returned to the fire, and boiled until half of the *lukk* solution remains. It is then removed. Write with it. If it is desired that it remain dissolved, a piece of hard white sugar[214] is added to it. If it is desired dry, it is placed in the shade protected from dust. When it is dry, it is removed and used for that which is desired. The *lukk* is broken into pieces and powdered like the crumbled chick pea.[215] It is washed with water and put in a thick filter. The water is boiled vigorously. While it is in the filter,[216] hot water is poured on it so that its color, red, will flow from the filter. The filtrate is boiled until it is decreased by two-thirds. Then dissolved gum is melted in it. Write with it. It comes out well.

Another color is the gleaming red ruby. Three *raṭls* of Carthamus[217] are pulverized a day in the sun. It is then pulverized and sieved[218] with a sieve which is larger than the flour sieve and smaller than a regular sieve. Then it is hung in a cloth above a wide filter on

---

209 *furūzjī*. Turquoise (*CAS*, 82) was known to the Babylonians. It is one of the stones in Aristotle's lapidary (11). The Arabic is probably from the Persian. The turquoise is mentioned in Diosc. (V: 136) as θυίτης λίθος.

210 *qirsh*. A coppery color.

211 For rubrica, *cf.* Ruska, J., Studien zu den chemisch-technischen Rezeptsammlungen des Liber Sacerdotum, *Quellen u. Stud. z. Gesch. d. Naturwiss. u. d. Medizin* **5**: 90, 1936.

212 *naranj*. Probably from the Persian. I. B. (2204) says that it is good for the heart and acts against nausea, according to ibn al-ᶜAwwām. The rind, pulp, and oil of the fruit are used. The pulp is used for the stomach. The rind macerated in olive oil is good for the bite of the scorpion and other venomous animals. The orange is *Citrus sinensis* Osb.

213 *yāqūt*. According to Aristotle's lapidary (3), there are three main types, the red, yellow, and dark blue. The red is supposed to be the noblest, most costly, and most useful. The ruby is not found in Dioscorides or Galen. The ruby is mentioned in Rasaratnasamuchchaya as a gem. Gems were regarded as agencies which helped the fixation or coagulation of mercury (Ray, 177). I. B. (2299) mentions that ibn Sīnā used *yāqūt* as an antidote for poisons. Al-Rāzī used it as an anticoagulant.

214 *sukkar ṭabarzad*. Al-Razzāq (829) wrote that this was crystallized sugar, perhaps what we know today as rock sugar. He says that this sugar is equivalent to the candy type. Al-Kindī used crystal sugar in his perfumery (nos. 21, 22). Al-Rāzī (48) claimed that the word *ṭabar-zad* meant "bile-split." I. B. (1198) states that *sukkar ṭabarzad* is not an emollient while the *sulaimānī* and *fānīd* types are. Al-Rāzī claimed that the sugar (*fānīd*) from Sejistān is a laxative and carminative. *Cf.* also I. B. (1199) for *sukkar al-ᶜashr*.

215 I.e. its size.

216 *rāwūq*. From Ar. *rawwaqa* "to clarify." *Raiq* = the best part of a thing. The Sum. *ŠIM* = Akk. *riqqu* or *rīqu*, a cognate of *rawwaqa*. In other words, the gum resin (in Akk.) is related to something filtered. *Riqqu* then represents the gum or resin which has filtered forth from trees. (*Cf. DAB*, 336–337).

217 ᶜ*aṣfūr*. A species of Carthamus. Maim. (300) gave the synonyms as *al-marīq*, *al-iḥrīd*, *bahram*, *bahramān*, and *al-sukarī*. The seed is *al-qirṭini*. Carthamus was used for textile dyeing in ancient Egypt (Lucas, 175). Diosc. (IV: 188) says that the seed is red and white. Two species, *Carthamus lanatus* L., the cultivated, and *Carthamus tinctorius* L., the wild, are known (*cf.* I. B. 1791). Ducros (161, 180) states that it is still sold today as a dye, carminative, and laxative. *Cf.* also *Tuḥfat al-aḥbāb* (348). ᶜ*aṣfūr* is sometimes called safflower. It is today cultivated in Syria, Iran, and Afghanistan as a field crop for its red florets which are used as a dyestuff and cosmetic (Hooper, 84).

218 *gharbal* = to sieve. gharbāl = a sieve. This passage indicates that the flour sieve possessed holes of a standard size. The sieve to be used is to have larger holes which are, however, smaller than those of the common size.

a support used by the dyers. While it is hanging, about sixty *raṭls* of water are poured over it. Let it drip into a basin[219] so that not a drop of water remains in it. Then that water with which it was washed is poured over it. The Carthamus in its cloth is then taken. For it, ten dirhams of black alum of the dyers[220] are pounded. It is sprinkled over it more than once while it is being rubbed well with the hands. This is done each time until the palms are dyed with its redness. It is then hung again and twenty *raṭls* of pure water poured on it. (**38**) It is allowed to drip until not a bit of the water remains. What has dripped is the essence of the needed Carthamus. With it are mixed a *raṭl* of wine vinegar and some water of gum. It is used the same day. Its color comes out well. Nothing else mixes with it. It can be applied to gold, silver, and tin.[221] It comes out well. When used on paper or parchment, it comes out a wonderful red color.

Another color is blood of the gazelle. Gallnut is taken, cleaned, and moistened with water. It is cooked and its juice used. Carthamus seed is cooked with water of perfume. It is all put with a dirham of Kufic ink, one-half dirham of alum, and one-half dirham of gum arabic. Write with it.

Another color is apricot. Yellow, feathered arsenic is pulverized well on a stone. It is then used. If an aroma of *galia muscata* is desired, the red arsenic is pulverized alone or red lead is added. It is broken up in pulverization. It is then removed. Write with it on anything which is desired.

Another color of it. The best Tūsi light indigo is pulverized with water. It is then washed well until it is clear and the soft parts remain. It resembles the *lāzward*. It is dried. When needed, it is melted in gum water. Use it.

Another color is pistachio. Yellow pulverized arsenic is kneaded with a little of the worked-with indigo and water of gum. A desired amount of indigo is added, step by step, until the goal is reached.

Another color is brown. A little red earth[222] is added to a needed quantity of *bauraq*. (**39**) The procedure is according to the first recipe. It comes out nicely.

Another color is that of dates beginning to ripen. Three ounces of sapanwood are pulverized well together with an ounce of Yemenite alum. Enough water is poured over it to cover it. It is boiled until the gum of the sapanwood comes out. It is then filtered. The filtrate is mixed with water of red *lukk* and with three dirhams of pulverized glue. Write with it at the time. It comes out wonderfully.

Another color is light. A desired amount of yellow pulverized arsenic is mixed with water of gallnut and water of gum. They are pulverized well, then dried. Then a part of it is pulverized with its sixth of the best indigo, with water of the rocket plant, and water of the green coriander after its water has been clarified. Use it on what you wish.

Another color is marble-like[223] white. White *bauraq* in which there is no blue is well pulverized on a stone. It is then sieved with a silk cloth. It is pulverized again. Water is dripped on it. It is dried and sieved. It is removed and water of gum mixed with it. It is kneaded vigorously and used for writing what is desired.

Another color is of the colors of the wild. One part of red earth is taken and the same amount of white lead, and also a little yellow arsenic. This color is of the fresh wild animal. If the color of the lion is wanted, then a little bit of *lāzward* is added. It comes out as described. If the color of the falcon is desired, then the required amount of *bauraq* is put on a small quantity of yellow arsenic and (**40**) as much as a fourth of arsenic of the black type. It may be increased in the measure desired. It comes out beautifully.

Another color is Chinese galanga[224] root. Sal ammoniac is well pulverized, kneaded with water of gum, and then removed. It comes out nicely. It is good for holy books.

Another color is pomegranate blossoms. What is necessary of the root of the saffron is mixed with the

---

[219] *ijjānah*.

[220] *shabb al-ṣabāghīn al-aswad*. This is probably alum mixed with green vitriol. It was used to dye leather and cloth black. *Cf.* I. B. (1279) for *shabb*. The white *shabb* or "white vitriol" is the best-known alum (Maim., 368). See also Alaune (121).

[221] *qazdīr*. Probably from Persian. More frequently tin is *al-qalaʿī* (Alaune, 110; al-Rāzī, 105).

[222] *maghrah* or *mughrah*. Red and yellow ochres were known in Babylonia (*CAS*, 123), in dyeing and painting. These were probably among the earliest coloring agents known. It is likely that impure ferric oxide in the form of some earth was used to color glass red. Red earths are described in Diosc. (V: 96) where *rubrica sinopica* (μίλτος Σινωπική) is the best; it comes from Cappadocia. "It is purified and brought to the city of Sinape where it is sold." Diosc. states that it is good for plasters, for the belly, and liver. The other is the artificial type (τεκτονική) obtained by calcination of yellow ochre. In ancient India, Rasaratnasamuchchaya gives two kinds of ochre (*gairika*), *pashana gairika*—hard and copper colored, and *swarna gairika*—golden yellow colored (Ray, 173). Maim. (238) gives the synonyms of *mughrah* as *al-mishq*, *al-ṭīn al-aḥmar*, and *urtukiz* (Turkish). In the *Tuḥfat al-aḥbāb* (196, 197, 198), various earths are discussed without clear distinctions. Ducros (15) states that *ṭīn armanī* is also red ochre sold in the bazaars as a tonic, dessicative, styptic, and for dysentery.

[223] *rukhām*. Marble.

[224] *khulanjī ṣīnī*. The *Tuḥfat al-aḥbāb* (411) gives the synonyms for *khūlanjān* as *khūdenjāl* in Morocco. It is stated to be the root of *Alpinia officinarum* Hance. The Persian is *khalanjān* (Steingass) which came from Chinese *kawliang-chang*. Thompson (*DAB*, 10, 11) thought that *šamurbatu* and *šamurpatu* in Akk. referred to galingale. This is probably the greater galingale called *asal* in Arabic and *'ārbāthā* in Syriac. However, *khūlanjān* is probably the lesser galingale. The lesser today comes from Indochina and China (Ducros, 100). It is particularly indigenous to the Chinese island of Hainan and is cultivated on the neighboring coast of Kwangtung and in Siam. Hooper (133) states that it is an ancient spice and medicine of the East. Bayān (26) mentions galingale in an electuary of musk which is useful for palpitation of the heart, fainting, and cold of the liver and intestines. It is employed as an aromatic, carminative, and aphrodisiac. It is also used as a condiment and stomachic. It cannot be the *khalenj* since this is indigenous mainly to Andalusia (I. B. 814).

same quantity of sharp vinegar. It is left to settle. It is well purified and mixed with hairy saffron boiled with pulverized gum arabic. It is used for what is desired.

Another color is violet. A little indigo is added to the mentioned saffron until it is pleasing. It comes out violet. It is mixed with water of gum. If it is too red, a little indigo is added to it. Write with it.

Another color is *lāzward*. Saffron is pulverized, then boiled with water of gum until its dye comes out in the water. Then sieved pulverized indigo is thrown into it and some red lead. It remains overnight and is then filtered in the morning. It is put on the fire and with it its fifth of gum arabic and its tenth of fish glue. It is boiled until it is dissolved and reddened. Write with it. It comes out beautifully like *lāzward*.

Another color is yellow. Monk's arsenic[225] is pulverized on a clean stone so well that it is not affected by the hand stone. It is put in the sweet water. Some saffron and gum arabic are thrown on it and pulverized. It is put aside in *līq*.

Another type. Shining red arsenic is pulverized well with water. If desired, saffron is dissolved in it. If desired, it is left as it is. (**41**) Then it is removed, in a *līq* which is in a glass vessel. Write with it after gum has been added. If desired, cinnabar is added to saffron. Use it.

Another color is green. Yellow monk's arsenic, which is better, is pulverized well with water on a stone. Good indigo is thrown on the arsenic. This is then pulverized. If pistachio is desired, then the amount of indigo is not increased. If myrtle is desired, or verdigris, or *misīnā*,[226] it is experimented[227] with by increasing the indigo. It is filtered, then put into the *līq*. Write with it. It comes out beautifully.

Another color is the white of fat. *Bauraq* is well pulverized with water. A small bit of dissolved *lukk* is thrown on it and then it is pulverized. It comes out fat-white. If a rosy color is desired, *lukk* is added. If purple is desired, indigo is added to it together with water of gum. It is removed in *līq*.

Another color is blue. *Bauraq* is pulverized well. A small amount of indigo is thrown on it. It is pulverized. Use it. If a darker blue than that is desired, then indigo and gum arabic are added to it. It is removed. If various shades of color are desired, these are obtained by increasing or decreasing the indigo.

Another color is the *rīhān*[228] color. Three dirhams of indigo are pulverized on a stone until it is a paste. On it is thrown a dirham of verdigris. It is pulverized until its color is pleasing. Write with it.

[225] Yellow monk's arsenic is probably a purer yellow sulphide of arsenic used by scribal monks in manufacturing their ink.

[226] *misīnā* may mean a green color produced by copper as a type of impure verdigris. *Cf.* Diosc. (V: 100) μιον.

[227] *jariba* is to experiment or to try out.

[228] *rīhān* is probably a shade of purple.

[228a] *Cf.* Appendix, chap. 5.

SEVENTH CHAPTER

ON THE WRITING ART WITH GOLD, SILVER, COPPER, TIN, AND THEIR SUBSTITUTES

Section on solution of the gold. (**42**) Pure gold is beaten to a thin leaf. It is then cut up into small pieces. On it is poured borax.[229] The fire is then introduced and blown on it until melted. It is then rubbed with a stone until it becomes like butter. It is then gathered and pressed until the liquid comes out and the gold remains. It is then returned to the rubbing stone and rubbed again with water of alum[230] used for wool and *Andarānī* salt,[231] table salt, and Greek vitriol. If its color is pleasing, then it is completed. Write with it like ink. It is good.

Preparation for goldwriting. A sheet of gold is put on a rubbing stone and good wine vinegar poured on it. It is pulverized three days. Then it is washed finely with water. Write with it. If desired, water of tragacanth can be used in place of vinegar. Water is poured on it. It is kept wet a day and a night until it becomes like honey in appearance. Then the powdered gold is washed. On it is thrown a measure of tragacanth such that it flows. Write with it.

Preparation for gold writing. What is desired is filed with a fine file. The filings are poured in a glass pot. On it is poured black *tūr*[232] filings. It turns black. It is left in it twenty one days in a place where there is no dust, no wind, and no sun. It dissolves. If it is desired to write with it, then red alum[233] is moistened in sweet water a day until the night. Then the pen is taken, dipped into water of alum and then into the gold. The supply comes from it. Write well with it.

For a similar writing, it is well filed. Then an equal amount of mercury[234] is added. It is then pulverized on

[229] *bauraq*. In this context, the meaning is borax which acts as a flux. In most cases, however, *bauraq* indicates a type of impure natron. Borax is called *tinkār* in the *Tuḥfat al-aḥbāb* (401), also *liḥām al-dhahab*, *liṣāq al-dhahab*, and *milḥ al-ṣāgha*. *Cf.* Aristotle's lapidary (46) for use with metals. *Cf.* I. B. (431).

[230] Water of alum was a well known mordant in Babylonia for dyeing of wool. (Levey, 163) It was also known as a flux (Levey, 160) and as a tanning and washing agent (Levey, 70, 122, 161) Although alum was mined in Egypt, there is no evidence that it was employed there in ancient times (Lucas, 291–293).

[231] *Andarānī* salt is a crystalline salt (probably a purer sodium chloride) found at Andaran in Persia. Some say it comes from Andera, a village in Syria. According to Alaune (121 ff.), "the best salt is that from Andaran; then comes the table salt, then the Indian salt which is red, and then the bitter." Salt is supposed to whiten the metals and to purify them. It also dissolves them and "strengthens the spirits." In salt lies a great secret. It is the "soap of the wise." *Cf.* I. B. (2164).

[232] *tūr*. This reading is uncertain.

[233] Red alum was probably alum with the red impurity of iron compounds. No doubt the mixture was found in many regions.

[234] *zaibaq*. Mercury was known to the Babylonians as was cinnabar, in fact, mercury was called *IM.KA L.GUG* in Sumerian, meaning "distillate of the red." Diosc. (V: 95) states it was kept in glass, leaden, tin, or silver vessels because of its corrosive property. An antidote for mercury is milk or wine with wormwood. In Arabic times, a good description of the properties of mercury is given in Alaune (90 ff.) Aside from perfumery materials, cinna-

a stone three days, then put in a thick cloth until (**43**) there comes out what is in it of the mercury. What remains of it flies away in the flame of the fire. Gum in the necessary amount is then added. Write with it.

Another preparation for gold. Yellow sulphur,[235] white alum, and wax,[236] in equal amounts, are melted, then all of it pulverized with one part of yellow arsenic, a half part of saffron, three parts of gum, and dissolved mica until its pulverization is well done. Write with it.

Section on writing with silver. It is thinned out as a sheet as much as possible. It is cut into little pieces and put into an iron spoon[237] on a fire. It is heated until it is warm. Then its weight of good mercury is added. It is pulverized with a clay pot handle and rubbed with it vigorously until all its blackness comes out and the water comes out clear when it is poured in. It is put in a thick cloth and gum arabic is put on it. Write with it.

Another preparation. Filings of silver are pulverized with distilled wine vinegar three days. It is then dried and pulverized again with the distillate until it comes out like clay. It is washed from the vinegar until its sourness[238] disappears. Gum is added. Write with it.

Another preparation. Four parts of white Indian tin[239] are melted. An equal amount of mercury is thrown on it. When mixed, it is pulverized on the rubbing stone until it becomes like collyrium. It is washed with water and salt with care until its blackness and impurity come out. The gum of tragacanth and gum arabic are added. Write with it on whatever is desired with a quill polished with a shell. Write with the quill.

Preparation of one similar to silver. Unslaked lime[240] upon which no water has been poured is pulverized. (**44**) On it is thrown thin melted glue. It is kneaded and made into flat cakes and dried. It is used.

Preparation of an ink[241] similar to the Chinese. Ten ounces of sieved soot of chick pea and three ounces of pulverized acacia are mixed together with pulverization. Water of sorrel is poured on it, and also five dirhams of salt and three dirhams of blue vitriol. All of it is well pulverized and left until dry. It becomes a powder which may then be sprinkled. Then thirty dirhams of gum arabic and three dirhams of gum tragacanth are powderized for it. It is kneaded with a little water. It is made into flat cakes, dried, and put in the shade. Water of gum is added to it when needed after pulverization. Use it.

Preparation of ink which takes the place of the chick pea. The surface of paper is brought close to the fire. It is covered by a wooden plate so that its strength does not disappear. Then its blackness drops off. This burned material is then pulverized. The water is extracted from sorrel leaves which have no veins. Then gum and salts, in the amounts necessary, are added. It is then boiled on the fire until it is dissolved. The froth is removed bit by bit. It is all put together. It is put in a flat dish which is stationary. The ashes are dissolved in it. It is then kneaded continuously with the palm until it is completed. The flat dish surrounds it. It is rubbed on warm ashes which had come about in the morning. It is then removed and used. It comes out well.

Preparation of another ink. A wick immersed in

---

bar is the first substance for which there is evidence for distillation. Mercury, according to the evidence, may have been the first substance sublimed for the purpose of purification. I. B. (1143) quotes various authors who agree that mercury is lethal to humans and rats. Its synonyms, according to Maim. (139), were *zāwūq* and in the Maghrib, *al-zawq*. *Zaibaq* comes from the Persian *jīva*.

[235] *kibrīt*. Aristotle's lapidary (26) discusses the various types of sulphur, red, yellow, white, and one which has all these colors. It is stated that all sulphurs are effective for sores. The water of sulphur is supposed to be pharmacologically effective for fevers, bile disorders, and swellings. "When sulphur is brought together with a mineral in the fire, then it burns up." Diosc. (V: 107) wrote that sulphur (θεῖον) was effective for scorpion bite, itching, for catarrh, hardness of hearing, and for bruises of the ear. It is also good for coughs and leprosy. Jābir states in *Book of the Garden* that arsenic takes the place of sulphur in the reddening, and sulphur takes the place of arsenic in whitening." Sulphur is still used in the Near East (Ducros, 196) for ulcers, itch, ringworm, herpes, and in industry for jewelry. See note 117.

[236] *shamaᶜ*. The use of wax has a long history. In Babylonia (Levey, 85–96), wax was used for polishing furniture, in casting, to write upon, and for medical purposes. In the case of the latter, it was compounded with butter and fat of the ram's kidney, and other simples to be applied to the eyes. Wax was also used as a protective coating on copper to prevent oxidation. Wax (*dišpu* in Akk.) was also called in Ar. *mūm*. The latter is from the Persian meaning "beeswax." Diosc. (II: 83) considered wax, κηρός, to have a warming and mollifying effect, very good for dysentery. In India, in the eighteenth century, wax was used for plasters and for burns. Wax was obtained from many sources. I. B. (1340) states that wax was used in unguents in Arabic times.

[237] *mighrafah ḥadīd*. Equivalent to a deflagrating spoon.

[238] *ḥumuḍah*. sourness, acidity.

[239] *raṣāṣ qalaᶜī*. *Raṣāṣ* is either tin or lead in Turkish. *Raṣāṣ* is tin as is *qalaᶜī* in Arabic. *Cf.* Aristotle's lapidary (60) and Alaune (110 ff.). In the latter, there seems to be some confusion as to the meaning of *al-qalaᶜī*. In this text, it may indicate either tin or lead. This carries on the difficulty found in Sumerian and Akkadian texts in regard to this substance. The text must determine whether it is tin or lead.

In Babylonia, tin was alloyed with copper. From the third millennium on, there are many examples of Babylonian bronze. Tin was used in definite proportions with copper mainly to facilitate the casting process (Levey, 198 ff.). Thompson (*DACG*, 116) thought that lead was Akk. *abaru*, Sum. *A. BÁR* or *A.GÚG*. These were sometimes antimony. *AN.NA* in Sum. and *anaku* were tin and sometimes lead. Al-Rāzī differentiated between the two tins. Evidently, he considered one to be a kind of oxide (*raṣāṣ*, a burned substance) and the other, native tin (*qalaᶜī*) (*cf.* al-Rāzī, 105 ff.). Maim. (32) gives the synonyms of *ubbār* as *al-raṣāṣ*, *al-usrub*, *al-anuk*, and *al-qalaᶜī*. The editor of Maim., M. Meyerhof, calls these lead. Compare Ar. *ubbār* with Akk. *arabu*. I. B. (13, 1042) has *abbār* as lead and *raṣāṣ* as lead. It is certain that the confusion in terms arose from the lack of distinction of the properties of tin and lead.

[240] *jīr*. Known early in Babylonia, to Diosc. (V: 115), and India (Ainslie **1**: 194–195), the Muslims, therefore knew it well. Maim. gives the synonym as *nūr*. It is *al-kils* before contact with water (Maim., 260). The *Tuḥfat al-aḥbāb* (290) lists it.

[241] *midād*. Given as a remedy in I. B. (2098) for abscesses and as a desiccative. Ibn Sīnā also used it.

radish[242] oil is lit. (**45**) An iron pot is heated such that it is removed from the earth a distance so that only air can go under it. Remove the soot suspended from it; it acts as the soot of the chick pea.

Preparation of soot ink. Soot of the chick pea is sieved with a hair sieve. Two handfuls[243] of it are used and also five dirhams of Kufic ink. It is well pulverized. Then it is put with the soot in a pot or in a porcelain vessel. Gum arabic is soaked in water a day and a night. Then the sorrel is made fine; its water is taken and clarified. A solution of two parts of the water of gum and two parts of water of sorrel is poured on the soot bit by bit. It is collected by hand. When it is collected, it is stretched on a stone or board. It is left in the shade until it is dry. On its surface, a bit of water of gum is smeared. Then it is removed. If it is the Kufic ink as was first described, then it is pulverized, covered with water as has been described to you, and left a day and a night until it is settled. The water is taken from it, then fresh water is poured on it. This is done for three days until the water comes out clear. The residue remains in the lower part of the vessel. It is used with the soot and others.

Preparation of soot of the chick pea. The soot is dissolved in a flat dish. For every two ounces, there are pounded with it two dirhams of gum arabic and salt. The gum arabic is pounded and the water is extracted. This is not stopped until it becomes like clay. After it is dried it is removed and used.

Preparation of soot ink for paper. Light Persian ink is taken which when broken up does not show clay or earth in it. (**46**) It is soaked in water a day and a night. Then the water is poured out and it is dried. For it, a dirham of gum arabic is soaked and also five dirhams of ink. The ink is pulverized and kneaded with water of the gum. It is dried and put in the inkwell. Write with it. It comes out as a pure shining good ink from the first to the last.

Preparation of another. One part each of good Persian soot ink, gum arabic, and gallnut, and also one half part of burned paper are pounded, sieved, and kneaded with white of the egg. Balls are made of it and dried. It is put into the inkwell. Write with it. It is the best ink.

Preparation of a special paper ink. One part each of good Persian ink and gum arabic are pounded and kneaded with filtered water of gallnut. The latter is made from ten large gallnuts. They are pressed and one-half *raṭl* of water is poured on it. If it is desired to write, then it is diluted with water of gallnut every time the ink dries. Clear water is never added to it. It comes out well and cannot be erased. It does not disappear from the paper. If it is desired that flies do not fall into it, then fat of the colocynth[244] is added.

Preparation of ink of the seseli.[244a] The Arabic *sasālī* is burned well, then well pulverized on a stone or a rubbing stone. Gum or the fruit of the acacia is added. It is made into flat cakes. It comes out beautifully.

Preparation of a Kufic ink. (**47**) Rags of cloth are burned. It is covered with a tub after it has been burned. It is left a day and a night from the morning on. Then it is put in a hair sieve. It is rubbed with the hand like collyrium. Then a *raṭl*, which is the amount necessary, of gum acacia is moistened. Three-quarters of an ounce of the gum is used. When the gum is dissolved in the water, more is poured on it but not too much. It is pulverized in a mortar and then made into flat cakes. It is tried and found good.

Preparation of Kufic ink. A white clean linen cloth is stuffed into a new clay vessel never touched by fat. It is well luted with good clay until air cannot enter it. If wind enters it, it spoils. Dung is heaped on it, the fire is lit for a day and a night, then left until it grows cold. What is inside is then pounded, kneaded with milk, and collected. It is then prepared in the form of flat cakes, and dried in the shade. While kneading, moistened gum arabic is added. A very good ink results.

---

[242] *fujl*. In the Near East, in ancient times as today, the radish, *lapti ŠAR* and *puglu ŠAR*, was a ubiquitous vegetable (*DAB*, 51; Hooper, 163). The radish, *Raphanus sativus* L., is called *turb* in Pers., *tur* in Kurdish, and *mula* or *muro* in Hindi. (*cf.* Steingass). I. B. (1672) gives the uses of the radish, ῥαφανίς, by Diosc. (II: 112) and Arabic writers for the kidneys, as an aphrodisiac, for quartan fever, and as a stomachic. Too much radish provokes nausea and colic. The *fujl* is not the wild radish, *al-laḥlāḥ*, described by Maim. (217) as *al-fujl al-barrī* and *al-haḍamān*, and in Spanish *labashnā* or *labshanā*. The seeds which have the pungent taste of mustard are diuretic, laxative, and lithontriptic.

[243] *raḥah*. A measure of a handful.

[244] *ḥanẓal*. The pulp of the colocynth was used in Babylonia (Sum. = [šam]*UKUŠ.LI.LI.GI.ŠAR*, and [šam]*UKUŠ.TI.GIL.LA*, Akk. *lilikū*) as a hydragogue cathartic *DAB*, 85). In the Ebers papyrus (*ḏ* 3 *r.t.*), it is used in a remedy for the teeth (Grapow, 67). It was also used for intestinal worms, for ailments of the joints, for gout, rheumatism, sciatica, and for the eyes in India (Ainslie **1**: 83–86). Diosc. (IV: 176) used the colocynth, κολοκυνθίς, to assuage edemata and ulcer suppurations, for the ears, and the intestines. Al Kindī (39*a*) used *ḥanẓal* as an ersatz substance to raise the weight of saffron. *Aqrābādhīn* (122*a*) gives *ᶜalkam* a kind of colocynth, for the muscles, pain in the back, and the perspiration of women. Maim. (158) states that it is the same as *murrār al-ṣaḥrā'*, *al-ḥadaj*, *al-kabasa*, *ḥabbat al-habīd*, *al-sharā*. I. B. (714) quotes ibn Sīnā as stating that the oil in which the colocynth is boiled is good for injection into the ear. Many other Arabic authors wrote on the colocynth. The *Tuḥfat al-aḥbāb* (177) says that radish is called *taferzīzt* (or *tafersīt* in Berber). It is probably the *Citrullus colocynthis* Schrad. Ducros (93) lists the colocynth in use today as a cataplasm, astringent, and purgative. The *ḥanẓal*, according to Hooper (100), is grown in the Near East, India, and other regions today. In Persia, it is called *kharbūzah-rūbāh*, *ḥandhal* in Kurd., and *indrazana* in Hindi.

[244a] *kalkh*. The genus is uncertain but may be *Ferula*. Diosc. (III: 53, 54) wrote that *sasālī* has a drying faculty and is good for rheumatic eyes. In Gr., it is σέσελι. I. B. (1178) quotes al-Ghāfiqī to the effect that *sasālī* facilitates child delivery, dissolves concretions, dilates obstructions, is good for the stomach, kidneys, vessels, and combats flatulence. Maim. (285) among his words for *sasālyūs* gives *al-ḥarrā al-rūmī* "Greek mustard" *ṭaqāra*, *al-ṭarādaliyūn*, *al-kāshim al-rūmī*, "Greek lovage."

Preparation of ink for paper sheets. Good Persian ink and gum arabic, one part of each, and one-half part of burned paper are pulverized and kneaded with glair of egg, and sieved. Balls are made from it and it dries. It is put in the inkstand. Write with it for it is extra black.

Preparation of another ink. Eight *mithqals* of blue vitriol in a vessel is put in a furnace of the glass until it reddens. It is removed after three days. It is kneaded first with a very sour vinegar. (**48**) It is put in a furnace. When it is taken out, vinegar is poured on it, and also gum and gum arabic. Write with it.

Preparation of a color of ink called "crow." Three parts of Kufic ink, one part of *lāzward*, and one part of *lukk* are all mixed together and put in a flask. It is put in a *līq*. Write with it.

Preparation of another color of ink. One part of ink and nine parts of white lead are mixed and put in the *līq*. Write with it.

Preparation of another ink from it. Dried twigs of the palm[245] are cut up finger large, put in a clay pot, broken up, and introduced into the oven or furnace. It is taken out in the morning, pulverized, and kneaded with gum put into it. Write with it.

Preparation of a lead ink. White lead is kneaded with sour vinegar and put into a vessel which is then luted all around with clay of the art. It is put into the upper part of the glass furnace for three days. Then what is in it is removed and pulverized. Vinegar and a bit of gum are poured over it. Write with it.

Preparation of an ink with glass.[246] A desired amount of glass is well pulverized and soaked with water until it becomes black and the glass is purified. It is then dried and put in a wide-mouthed flask. A bit of pure gum arabic and wine vinegar are mixed well with it. It is hung in the sun for seven days in the summer in the strongest part of the heat. It is stirred every day. Every time it dries, it is soaked with wine vinegar. If it is desired to write with it, the pen is inserted in the flask. (**49**) Then the latter is covered to protect it from the dust as much as possible. The glass is stirred every day. The ink is taken out of the well with a copper pen to write. The mouth of the vessel must be covered from dust.

Preparation of another ink. Whichever soot is desired, the previous or any other, or any burned substance, is pulverized until a solid feeling is no longer there. It is sieved with a thick sieve. Then water of the leaf of the sorrel is pressed out and kneaded well with it until it comes out like soft dough. In every two ounces of ink, five dirhams of gum arabic are added and also a bit of mercury. It is pulverized with wine vinegar. Then it is put into a thick cloth. Write with it what is desired.

Preparation of writing with copper. Water of dissolved sumac is poured on filings of copper. It is pulverized three days, then dried and water of rocket added. It is pulverized until it becomes like fine dust. It is washed with water until it becomes yellow. Gum arabic is added to it. Write with it.

Preparation of another writing. Filings of copper are thrown into a glazed pot.[247] On it is poured *naft*[248] and gum arabic. It is placed in the sun for four days. Then it is powderized on a stone with water of alum, powdering it effectively. The filings are then pulverized and added to water of gum. Write. That is the writing with yellow copper by this method.

EIGHTH CHAPTER
ON RECORDING SECRETS

White vitriol[249] is used to write with. Then water of gallnut is smeared on. Or, water of gallnut is used to write with, and vitriol is smeared on. The well pulverized vitriol is sprinkled on and the writing appears. (**50**)

Description of writing with sal ammoniac. Sal ammoniac is dissolved in water. The water is not increased. It is left until it is dissolved. When it is dissolved and all of it becomes liquid, then it can be used to write with if desired on glazed paper sheet, on paper, or on parchment. It is left until it is dry. Then it is steamed with milk. If the soot touches it, then that writing is made to appear.

Description of writing with milk. Write with yoghurt on paper and send it to whom it is desired. The other man will sprinkle on it ashes of *qarāṭīs*: this is from the burning of paper; its ashes are sprinkled.

Description of another method. One-half mithqal of sal ammoniac is dissolved in a manner so that one can write with it. A dirham of *khawlān* which is the boxthorn[250] is thrown on it. It is left for twenty days; it is

[245] *nakhl*. Use of the various types of palm trees was made in very early times in the Near East by the Babylonians. One of the more common in Mesopotamia was the date palm, *ⁱˢGIŠIMMAR* in Sum., *gišimmaru* in Akk. Of the many synonyms, none is cognate to *nakhl*. In Persian, it is also *nakhl*. Dioscorides and al-Kindī used the palm but this is the only instance found where the twigs themselves are used. In al-Kindī (83*a*), only the marrow of the twig is used. The *nakhl* is most likely the *Phoenix dactylifera* L. The date palm today yields a very important crop in Mesopotamia (Hooper, 149–150) which is the largest date-growing district in the world. Over a hundred varieties of dates are known.

[246] Ground glass was used for its sparkle. The author has been informed by natives of Syria that it was still being used in the early part of the twentieth century.

[247] *barniyyah khaḍrā'*. A pot glazed on the inside.

[248] *naft abyaḍ*. White *naft*. Al-Rāzī (90) describes *naft* salt as being in hard black pieces without a shine but with an odor. Distilled *naft* is listed among oils in another section (al-Rāzī 156, 180). It is uncertain what *naft* indicates. Lane shows that it was used medically in colic, to open obstructions, and, when used in a suppository, to kill worms in the womb. Bayān used *naft* in a confection for phlegm and old sores. In the present day, *naft* signifies naphtha and petroleum.

[249] Probably a purer alum. A pure alum used in the Ebers papyrus for the eyes was *'ibnw* (Grapow, 59).

[250] *ḥuḍaḍ* or *ḥuḍuḍ* is juice of the boxthorn. According to Maim. (148), "*ḥuḍaḍ* is applied to the juice of the plant. It is *al-ḥuḍaḍ*

not exposed to the sun. It is then boiled vigorously. Then two dirhams of mercury are killed[251] and left for forty days. Ten dirhams of yoghurt are thrown on it. One writes with it a kind of writing which cannot be read except in the night and in darkness.

Description of another way. It is that two dirhams of goat yoghurt are taken and two dirhams of milk of the wild ass. All of it is thrown into five dirhams of grape juice where it remains ten days. It is dissolved in fifteen dirhams of camel's milk. It is that camel whose whiteness is inclined to redness. Then write with it. It cannot be read except in the light of the lamp. If a man has yellow jaundice and drinks one-half dirham, he recovers. It is also for those who have liver fever and similar to that.

Another type of it. The heart of the seed of the prune[252] is removed, pulverized, and sieved. Two dirhams of this are mixed with one dirham of *bauraq* and two dirhams of Greek gallnut. It is left for a month in the shade, then ten days in the sun. Then five dirhams of the milk of a woman are thrown on it. Write with it in a book. It cannot be read until powder of chalk is sprinkled upon it.

Another type of it. One and a half dirhams of gum arabic are mixed with cow's milk and one dirham of gum tragacanth. It is boiled but not too much. It is left for forty days. Then three dirhams of water are put on it. Write with it. It cannot be read until ashes are put on it.

NINTH CHAPTER

ON THE WORK IN WHICH THE WRITING IS ERADICATED FROM MANUSCRIPTS AND PARCHMENT, FROM PAPER AND BOUND BOOKS

One part each is taken of yellow Yemenite alum, false bdellium,[253] alum of the Carthamus, and white sulphur. It is pounded well and soaked in wine vinegar, then pulverized until it becomes like fat. It is made into the shape of an acorn. Whatever desired may be rubbed with it in order to make it look white.

Another way to eradicate writing. One part each of white and blue bdellium and yellow sulphur are taken and pulverized with wine vinegar. It is shaped into an acorn. With it ink is erased from paper. It comes out.

(52) Another preparation with which gall ink is removed from parchment. Water of soap is mixed with a like amount of vinegar. It is distilled. When it is written over the letters, then it removes the ink from paper and parchment. Like that is the water of the distilled onion[254] with water of the distilled soap.[255] The operation is conducted in the same manner.

Another type is when the ink for paper and parchment is peeled off and its traces are removed. White *iqlīmiyā*[256] scum is taken from a melting metal, pul-

or *fīlazahraj* (Pers.) and the Greek name of this juice is λύκιον; it is the collyrium of *Khawlān* (*kuḥl Khawlān*)." The Persian name comes from *fīl zahra* "bile of the elephant." In Babylonia, lycium was probably *Ú.GIR* (in Sum.) and *ašagu* or *iṭṭitu* (in Akk.). In Hebrew, it is *atādh*. Its shoots, roots, seeds, tops, and powder are found in Babylonian texts for a blow, toothache, excessive salivation, and difficulty in menstruation (*DAB*, 182 ff.). The Arabic *ᶜausaj* is the boxthorn and its uses are those of the *ašagu* described. Diosc. (I: 100) prescribed lycium for the eyes, fumigation against venomous beasts, and as a tonic. Leclerc (I. B., 680) says that *ḥuḍaḍ* is *Rhamnus infectoria* L. Al-Kindī (41) used it in a musk recipe. The *Tuḥfat al-aḥbāb* declares that *ḥuḍaḍ* "is *al-khawlān* of Mecca." The Berbers call it *arghīs*. *Khawlān*, according to Freytag (I: 538), is the name of an Arabic tribe in Yemen where the collyrium was originally known. Today *ḥuḍaḍ al-yamānī* is still known mainly in incense and used in witchcraft (Ducros, 168).

[251] To do away with many properties (both physical and chemical) of mercury. Nagarjuna said, "Killed mercury is that which does not show signs of fluidity, mobility, and luster" (Ray, 134). Kakachandesvarimata Tantra states that mercury is rubbed with ingredients of the *vida* and roasted in a closed crucible. The ingredients with which the mercury is roasted may include vegetable matter, sulphur, and other substances. (*Cf.* al-Rāzī, 102–104; Aristotle's lapidary, 61.).

[252] *ijjāṣ*. The prune has been well known for thousands of years in the Near East. Maim. (13) gives the synonyms for *ijjāṣ* as *al-burqūq* in the Maghrib, *ᶜuyūn al-baqar* in Spain, and also *al-shāhaluk* and *al-shāhalūj*. The latter are Persian. I. B. (21) describes the use of prunes in Arabic writers as a laxative, for bilious fevers, and as a stomachic. The *Tuḥfat al-aḥbāb* (45) mentions *ijjāṣ*.

[253] *muql*. Bdellium was probably known in Eighteenth-Dynasty Egypt (Lucas, 373). Maim. (230) has a good description stating, "This name is applied to the resin of a tree . . . called *al-kur*. . . . The resin of this tree is often found in medical books; it is called *al-muql al-azraq* (blue bdellium) and *muql-yahūd* (Jewish bdellium)." It was known to Diosc. (I: 67) as βδέλλιον. The Sansk. *guggula* may be related to the word *muql*. *Muql*, in Arabic times, was used for tumors, to cicatrise scrofula, as an aphrodisiac, and as an antidote for poisons, among others (I. B., 2157, 2158).

[254] *unṣul*. The onion, in Sum. *SÈ.SIKIL.ŠAR*, in Akk. *sikillum*, was used in Babylonia for dryness of the eyes. It is probably the *Allium Cepa* L. Diosc. (II: 151) wrote that onion was good for incipient cataract, for opening the orifices of blood vessels, for relieving hemorrhoids, and as a diuretic. The Greek name is σκίλλα. It is also called *baṣal al-fār*, *al-ishqīl*, and in Berber *uḥkāl*. (Maim., 60). The *Tuḥfat al-aḥbāb* (31, 303) mentions *baṣal al-khinzir*, "onion of the pig." Al-Ghāfiqī (134) says that onion is injurious to the mind. When cooked, it is an aphrodisiac, and, eaten raw, it checks the harmfulness of different kinds of water.

[255] *ṣābūn*. Soap is found mainly in medical works in early Mesopotamian and Egyptian literature. The oldest recipe for soap made from water, alkali, and oil, is given in a text from the Third Dynasty of Ur (*ca.* 2200 B.C.). A Sumerian medical tablet of this period gives another preparation from *Salicorna fruticosa* L. (an alkaline plant) and cassia oil among other substances. This was sprinkled upon the patient followed by a rubbing with tree oil. A true soap, using caustic alkali, was not produced in antiquity. The early soaps resembled those of the present day known as cold or semi-boiled soaps. The impure carbonates of sodium and potassium from plants were used to make these soaps. It is still done today in remote regions (Levey, 125–129).

[256] *Iqlīmiyā* comes from the Greek καδμεία; the Latin is *calamina* (al-Rāzī, 50). The meaning is uncertain. Maim. (342) wrote that it is the scoria of the metal in fusion. A description of gold *iqlīmiyā* is given in Aristotle's lapidary (138), "When gold is added to another metal, then placed in the fire, its substance is purified and it drives another stone high, mixed with black, in part made up of the dye of glass. It is the stone, *iqlīmiyā* of gold."

verized and soaked with acid of the citron.[257] Then whatever is desired is rubbed with it. It comes out.

Description of the disappearance of gall ink from parchment and paper. Wool is immersed in yoghurt. It is used to rub the writing by hand. Then a little bit of kneaded salt is used. It then disappears.

Preparation of another eraser for paper. One part each of *bauraq*, gum arabic, and white sulphur are finely pulverized and made into a ball. It is dried in the shade. If it is needed, a little water is put on the end of the pen and the material is rubbed on the writing. (**53**) Write above it whatever is desired.

Preparation of another eraser for paper and parchment. It is effective. Into a pot which is glazed on the inside, there is put a *raṭl* of *sabkhī*[258] salt or *Andarānī* salt or some other whatever it is. After two dirhams and no more are poured on the salt, then it is put into the distillatory. It is distilled until the dripping stops. The distillate is kept so that nothing enters it to take its strength. What remains from the salt which was not distilled is put aside and returned. In the pot is put one half *raṭl* of another fresh salt. On it is poured the water of the first distillate from the distilled salt. The water is separated after it has been kept from the air. The remainder of the salt in the pot is thrown away. Another new *raṭl* of salt is returned. On it is poured the distillate. It is distilled. This is done seven times. At the end of the seventh time, it comes out white. The pen is dipped into this water which is written over the original letters until a trace of these cannot be found on the paper. It comes off at the same time until no trace is noticed. It uproots all the dyes of the cloth and tanning.

TENTH CHAPTER

ON THE WORK OF GLUE AND SNAILS, SOLUTION OF FISH GLUE, ADHERENCE OF GOLD AND SILVER, DESCRIPTION OF POLISHING AND ITS POLISH, WRITING INSTRUMENTS OF HAIR AND FEATHER, ALL OF THE INSTRUMENTS FOR GOLD WITHOUT WHICH GOLD CANNOT BE WORKED SO THAT IT DOES NOT DISAPPEAR[259]

Clean white fish glue is crumbled and soaked a night in sweet water. In the morning the water is poured off it. It is kneaded with the hand until it whitens and becomes like wax. It is put into a copper vessel[260] whose poison has been removed. It is raised onto a low fire until it is melted. It is filtered through a cloth. Use it.

Description of the work in preparation of snail glue.[261] It must not disappear. What is desired, five handfuls of the desert snail, is put into an iron mortar. It is made very fine, then put into a lead pot[262] all day on the fire. (**54**) Water is sprinkled on it throughout the whole day little by little so that it will not burn and as much as one desires of oil until it is ready. When the cooking is completed and it has become viscous, then this is the best glue without which the best work of gold on paper cannot be accomplished. It is the best glue for pictures since it does not ever break off and it remains whole. A *ḥuqq* container is used.

Description of the container[263] for dissolving the glue. It is round and thick on the bottom and has a rounded handle.

Description of solution of glue and how to make the gold adhere. White, sticky, easily broken, fish glue is cut up as small as possible and moistened in sweet water a day until it is soaked and has become soft. When it is moistened, it is gathered, and rubbed finely and gently until it is soft. It is then gathered and put into a vessel. On it is poured sweet water. It is raised on a low fire until it is dissolved. When dissolved, a little puverized saffron is added in a quantity that changes its color. Then it is filtered with a thin clean cloth. When the weather is hot or cold, a fire should be present since it solidifies quickly. When it is thickened, it is placed on the fire until it melts. When you write with it whatever is desired, the excessively red, beaten Ibriz[264] gold leaf is pressed on that glue a day. It is not delayed more than that. (**55**) If the gold does not stick properly with the glue, then the gold is heated on the fire and the alum shaken from it so that the whiteness cannot change. When it is pressed, it is left for two days and polished with a stone.[265] Then *kuḥl*[266] is put on it.

I. B. (1826), under *qalīmīya*, says that Galen claimed that this substance was formed in the copper furnaces. Diosc. (V: 74) states that *cadmia* comes from the soot sticking to the sides and top of the furnace when brass is heated red-hot. This is also true for heating of silver.

[257] *utrujj*, from Pers. *turunj* which is from Sansk. *mātulunga*. It is probably *Citrus medica* Risso or some other species. Maim. (1) called it "medicinal apple." Diosc. (I: 115) knew this as did the ancient Babylonians (in Sum. *A.AM*, in Akk. *ildaqqu* and *adaru*) and Hebrews *ethrōg*, used in Hebrew rituals. In the Arabic period, I. B. (16) wrote that the citron is common and that it is good for febrile palpitations, as an antidote for scorpion bite, for yellow in the eyes, for the stomach and the viscera.

[258] *sabkhī* salt is a paste form of salt (I. B., 2168, 2164).

[259] A difficulty with which the book ornamenter was concerned was the falling off of gold leaf and paint. The author is obviously concerned with this here as well as throughout his discussion. *Cf.* Appendix, chap. 5, notes *a*, *b*.

[260] *inā' nuḥās*. It is still held by many that containers made of metals impart a poison to the contents. The text, unfortunately, does not relate the process of removing the poison. This idea, by the way, is a very old one, going back thousands of years to the period succeeding the Stone Age. Later reflections of this idea are to be found in Hammurabi's Code and in the Old Testament where surgical knives of stone were favored over metal ones.

[261] *ghirā' al-ḥalazūn*.

[262] *qidr raṣāṣ*.

[263] *ḥuqqah*. A receptacle cut from wood or ivory.

[264] A very pure gold.

[265] *maṣāqil*.

[266] *kuḥl*. Most of the time in the Arabic period, it was antimony trisulphide (Sb_2S_3), stibnite, with many and varying types of impurities. This was not only true with the Arabs but with the Babylonians, Egyptians, and Greeks before them. I. B. (18, 1898, 1899, 1900) states that *kuḥl* or *ithmad*, as known by the Muslims, was a collyrium for the eyes. It stops uterine hemorrhage as a suppository and it dessicates ulcers. One of its main uses was as a cosmetic for making the eyebrows and lashes black. Some believe that *kuḥl* was native antimony (Maim. 27). In ancient times, antimony was frequently confused with lead. *Al-kuḥl*, in the mediaeval period, became identified with the quintessence of a substance, and thus spirits of wine or alcohol.

It is used in the polish with moisture on the middle finger between the gold letters, then with *kuḥl* after that.

List of the polishing agents of gold—the burnishers. For this art, there are three polishing agents of *jumāhun*[267] stone: the blue rounded feathered one; that which is rectangular in form, proportionate in the face which is at the head of the feathering since the sides are not used; and the third is small, pine tree in shape, with a proportionate face. The last is for the polishing of thin lines and its complications in fine work. Its fine edge is not pointed but it has a slight width in order that its purpose may be accomplished. A little piece is shaped for it, as much as the quantity of silver. For much gold, put the stone in the middle and lower it into the *lukk*. A cover is made for it either of silver or copper. It is made tight so that it will not shake because of the force of the work. For little gold, there are perpendicular pieces at the head of which are the stones. One performs with it as with the first. Then in the absence of *jumāhun*, onyx[268] is used in its place.[269]

Description of the polishing tablet. The polishing tablet for the gold is square, in the thickness of one finger. It is made from the willow[270] or the walnut for fineness in the work. (**56**) If it fails, then a tablet of another type of wood is used.

Description of the knife for gluing the gold leaf. It is an Indian knife whose length with its handle is the span of a king or two thirds. Its bared blade is wider than its handle to cut a leaf of gold or something else. The second side is indented; its middle section is wider than its extremity. It is good for softening of the pigments after their occurrence on the leaf and after they have dried.

Description of a sponge to push gold leaf in the pressing. A piece of sea sponge[271] is made round with scissors and put in a reed head. It is inserted with the fingers. Take some away from its head later.

Description of the quill pen for writing and so on. The part of the wings of eagles, thick with feathers, is taken. From it, the hard thick place is chosen and the quill plucked. The pen is cut off with the scissors since the knife does not do it entirely straight. A short cut is made for the pen. The fat is removed from it to make it thin. It is good for drawing and writing. The scissors used to cut off the feather pen are short at the head. The blade of the scissors should be thin.

Description of the brush pen. The hair of a weasel is taken and the thin part bundled, all of it to one side. Then Indian aloeswood[272] or sandalwood,[273] or something of ivory[274] or ebony[275] is thinned out for the hand

---

[267] *jumāhun* or *jumāhum*.

[268] *jaz*ᶜ or *jiz*ᶜ. Many types of onyx have been known from prehistoric times onward in Egypt, Mesopotamia, China, and India. Aristotle's lapidary (6) claimed that onyx came from two sources, China and the West. "It is black and white, yet not pitch black. It also may be mixed with green and yellow. Whoever uses it for a seal has many troubles and contends with bad dreams. . . . It is a very hard stone. . . . When it is pulverized and used to polish the Yaqut stone, it beautifies it and brings out its sparkle." From the description above, the onyx is not always the very black stone which is known by this name today.

[269] These were burnishers for application to metallic writing on paper and leather. Usually these had an agate or bloodstone set in a handle. This was still true in nineteenth century books on the subject. *Cf.* Delamotte, F., *Primer of illumination*, 40, London, 1860; Donlevy, A., *Practical hints on the art of illumination*, 14, New York, 1867.

[270] *ṣafṣāf*. The willow grown in the southern Babylonian marshes was the *Populos euphratica* Oliv., in Sum. ^{is}*ḪA.LU.ÚB*, in Akk. *ḫaluppu*. It was employed as a stomachic, in childbirth, for the feet, and as a poultice. It was a popular building wood. In protohistoric Egypt, it was used to make a knife handle (Lucas, 501). Diosc. (I: 104) used the *Salix* or willow, ἰτέα, *P. alba* L., for the astringent quality of its fruit, leaves, and bark. Some of it is good for the ears, gout, and the eyes. In the Arabic period (I. B., 815), the oil of *khilāf*, a species of *ṣafṣāf* was used in perfume. Whether the willow meant here is *Salix aegyptiaca* L. or some other species is uncertain. Maim. (393) states that *ṣafṣāf* is the same as *khilāf*, *sindār*, *gharab*, *sawḥar*, and *sālij*. The *Tuḥfat al-aḥbāb* (193, 412, 438) gives *khilāf* and *gharab* for the *ṣafṣāf*.

[271] *isfinj al-baḥr*. Diosc. (V: 120) used sponge, σπόγγος, for wounds, to repress edemata, and to open contracted ulcers. In India (Ainslie **1**: 401–402) it was considered to have antacid properties, to be a tonic and deobstruent, and to be good for herpetic eruptions and scrophulous affections. Galen used burned sponge (XII: 376) for hemorrhage after incisions. It was dipped into liquid pitch (al-Ghāfiqī 264). In Persian and Turkish, the word for sponge is pronounced *isfanj*. Maim. (5) wrote that it was also called *zabad al-baḥr* and *ghamām*. In the Maghrib it is known as *nashshāfa* or *ṣūfat al-baḥr*. I. B. (75) wrote that the Arabs thought the sponge should be used when fresh while it still retained properties of the sea. The *Tuḥfat al-aḥbāb* (42) calls it *al-jaffāfa*. The sponge is the *Euspongia officinalis* L. and others.

[272] ᶜ*ūd hindī*. Maim. (296) says that it is "that which the physicians call perfumed wood." It is the well known fumigation wood. It is called ᶜ*ūd al-nadd*, *al-*ᶜ*ūd al-khāmm*, *al-*ᶜ*ūd al-jaff*, *al-*ᶜ*ūd al-ṣanfī*, and *al-anjūj*. In ancient Greek it was ἀγάλλοχον. Diosc. (I: 22) used it, *Aloexylon agallochum* Lour. (?) when chewed for sweet breath. It was also employed as a stomachic, for dysentery, and for other ailments. In I. B. (1603) it is mentioned as a perfume and as a help to end urine incontinence. The exact species of aloeswood known to the ancients is still uncertain (Leclerc's note p. 485). The aloeswood is often found confounded with the legume sometimes bearing the same name. Al-Kindī used *agalloche* in his perfumery recipes. The *Tuḥfat al-aḥbāb* (368) states that there are thirteen types of aloeswood, the best being from India; then comes the *al-Samandūrī* (a village of Sofala of India near Bombay); and then the *al-qumārī* (from a region of India called Insulinde. The aloeswood is till sold today by druggists (Ducros, 167).

[273] *ṣandal*. From Persian *sandal* which is from the Sanskrit *candana*. Al-Kindī (113) used sandalwood in many of his perfumery recipes. It was also considered as a drug. Jābir (VI: 161*b*) discussed two types, the white and red sandalwoods. I. B. (1418) quotes an Arabic author as stating that this wood which comes from China is of three kinds—red, yellow, and white. The last, *Santalum album* L., is the most favored. Ibn Sīnā is quoted as having used it for the heart and stomach. The *Tuḥfat al-aḥbāb* (297) knew three types as did I. B. The white sandalwood tree is still grown in Persia and is called *sandal-i-safīd*. It is used in perfumery and in medicine for its stimulative and antiseptic action in the genito-urinary tract. Ducros (146) states that the white and red sandalwoods are still used in medicine.

[274] ᶜ*aj*.

[275] *abnūs*. Mentioned in Diosc. (I: 98), ἔβενος, it was a rare but fairly well known simple in the ancient materia medica. I. B.

so that it may be light for it. On its head is made a place for the attachment. The hair is prepared around the head after fish glue has been smeared on it. (**57**) This is to hold the hair fast. The thinnest pens have four hairs. They are made thinner[276] than that but this is stronger.[277] It is tied with a silk thread. Then fabricated Chinese oil is taken with buttery gum sandarac and is pulverized and sprinkled on the oil. The silk thread used to tie the hair is oiled and put in the sun until dry. It becomes like marble, hard and elegant. If it is washed with water, it is not changed and is not dissolved. Of it, there is made the thick and the thin. It is necessary to prepare two pens for every dye, thick and fine; for black, five of them, four for thin and one between thick and thin. If this hair fails, then its substitute is prepared as has been described. Hair of the cat or squirrel tail is tied in a bundle resembling the other in hardness and fineness of the head and shortness. It takes its place. Like the compass, it is essential that it be light in weight and delicate. It is demonstrated to be proper if, when it is opened a little and then closed a little, it covers without changing. Then it is good. One head of the pen is used to split it and tie the pen in a necessary manner. It is for delicate and fine work and what is wonderful in this art.

If gold is not available, then use a substitute in gilding. A *raṭl* of gallnut is dried in the sun, then powderized well. It is pulverized in a mortar and put in a cloth similar to a sieve. It is hung up. Sweet water is poured on it so that its purified water goes drop by drop without being changed. Then it is put in a woolen cloth. (**58**) The ends are brought over it. The cloth is pressed from the ends so that its water does not remain in it. If any of its water remains, then it will be corrupted. It is then spread on the leather. Four dirhams of dyers' alum[277a] are poured between the hands so that they are reddened, a red usual to the hand. It is returned to the cloth which is then spread out with the hand until all its parts are collected. Then sweet water is dropped on it, little by little, and the sides are taken care of so that it is filtered. The filtrate is collected. Its quantity is half of the residue or less than it. On the filtrate is poured water of sour pomegranates in the quantity of one ounce, or sharp distilled water. It is distilled; that which is distilled is purified every time. It is left until it is settled. It is repeated until the essence remains. When it is like the viscosity of honey, one-third of an ounce of red gum arabic is thrown on it. Then it is spread on a stone. When it is dry, it is removed when needed. When it is desired to use it, a solution is prepared with water and a little vinegar. One writes with it. It comes out a pretty ruby color.

If it is desired to inscribe on silver and on tin, then it comes out like gold. Half of the dry mass from the filtering of saffron is put in a copper pot. It is put on the fire until a third remains. It is tested with the pen on a finger nail. It becomes like honey and its color is golden; it is tested during its cooking so that it will not stay longer than necessary. The fire may be heightened or changed since the secret of this process is in the cooking. It is removed in a glass container. (**59**) When it is necessary for the operation of gluing it to silver or tin, then rub it on. It comes out like gold.

ELEVENTH CHAPTER

ON THE MANUFACTURE OF PAPER, IMPARTING OF DYES BY THE PEN, SOAKING OF PAPER, ITS DYES, AND ITS BEAUTIFICATION

Description of the manufacture of paper. The best white flax[278] is purified from its reed. It is moistened and combed until it softens. Then it is soaked in quicklime a night until morning. It is then rubbed with the hands and spread out in the sun until all of it dries in the daylight. It is then returned to water of quicklime, not the first water. It is so the next night until morning. It is then rubbed a night as in the first rubbing and spread out in the sun. This is done so three or five or seven days. If the water of the quicklime is changed twice a day, then it is better. If its whiteness is brought out, then cut it with the scissors little by little. It is then immersed in sweet water for seven days. The water is changed every day. When the quicklime has gone out from it, then it is pounded in a mortar very finely while it is moist. Then, nothing will be left of the lumps. Other water is put on it in a clean vessel. It is dissolved until it reaches a silky viscosity. Then it is introduced into the molds in the desired size. These are made from straw used for baskets, nails, and the walls are collapsible. Under it is an empty rib. The flax is beaten with the hand vigorously until it is mixed. Then it is thrown with the hand flat in the mold[279] so that it will not be thick in one place and thin in another. (**60**) When it is evened, then its water dries away. It is found proper in its mold. When the desired is attained, it is adjusted on a flat tablet. Then it is bound to a wall and straightened with the hand. It is left until it is dry. It separates and falls off. One may take a powder, shining white, pure chalk[280] and starch[281] in equal

(9) states that the Arabs used it for pustules and as a stomachic. The *Tuḥfat al-aḥbāb* (24) says it is *sāsim*. The Gunther edition of Diosc. gives ebony as *Diospyrus metanoxylon* Roxb. or *Ebenoxylon verum*. The ebony is known in many types. It is uncertain which type *abnūs* is.

[276] I.e. less than four hairs.

[277] The head of the pen holder is smeared with glue, then the hairs making up the fine brush are applied to the wet, glued end.

[277a] Probably the alum used as a mordant in dyeing (*cf.* Levey, 77).

[278] *qunnab*. *Cf.* Levey (87) for the uses of flax in Babylonia. The *Hibiscus cannabinus* L. (according to Hooper), called *jiljil* in Iraq and *palsan* in Hindi., is still grown today in India and near Basra for its fiber.

[279] *qālib*. A model or mold.

[280] *ḥuwwāra*. Known in Babylonia, it was also well known to Diosc. (V: 132) as good for eye ulcers. Diosc. called it γαλακτίτης λίθος.

[281] *nashā'*.

quantities. The powder and the starch are macerated in cold water until there is no lumpiness. It is heated to the boiling point. When it boils, it is filtered on that powder. It is stirred until it settles and it becomes a sheet. Then the sheet is drawn back and glazed with the hand, then put on a reed. When all the sheet is glazed, the sheet is dry. It is glazed[282] from the other side, then returned to a flat tablet. Water is sprinkled on it lightly. It is then gathered and stacked. It is polished as one with a cloth. Write on it.

Description of soaking of the paper. A very white kind of rice[283] is cooked vigorously in a pot[284] or in a glazed pan.[285] There is no fat in the pot. It is washed, then the water of the rice is filtered in a sieve or it is drawn through a clean cloth. It is then spread out on a clean cloth. It is so until it is dry. Some people cook the husks and take the water with which it was soaked. Some people wet tragacanth or soak it with starch. This is after it is boiled with water and soaked as described.

Description of beautifying the paper that has been tested. In a copper pot, ten *raṭls* of sweet water and good clean starch are cooked on the fire. It is boiled more than once until the water is diminished by two fingers or more. Then there is added a little saffron in a quantity to strengthen its (**61**) color or its purity. The solution is poured into a wide basin. The sheet is immersed in it lightly with care so that it is not torn. It is spread with a thin flax string in the shade. One must be careful that it is not réached by the sun else it will be spoiled. It is examined every hour with a turning over so that it will not stick. When dry, it is polished with glass burnishers on a board.

Another description of it. Old straw is moistened in water for three days. It is then boiled until a third of the water is lost. Starch, in the mentioned weight of the first description, is thrown into it. The operations are carried out as in the first procedure. It comes out improved for pen coloring and drawing.

Description of white writing on a black surface by pens. Sea reed growing in the meadows or the reed watered by rain or the one irrigated from time to time which is growing in free places, or the hanging vine—a large quantity of an arm's length is cut up after it is found smooth, hollow, and clear of knots. It is washed clean. Already wool alum has been dissolved in water. When its color is shown, then the pen is immersed in that water all over. It is thin and does not show too much on the body of the pen. The pen is dried in the sun. When the white dyes, it blackens. When it sticks to the pen quickly, then it is black. It is shiny and it stands out. Its whiteness is glittering and shiny—not at the same time. When it is rubbed, the second crust comes off. The black does not stick. There is no form of it as was first described. (**62**)

If the water of the alum on the pen dries, it is thought to be good. It is pulverized very thinly on a stone. After proper pulverization it is cooked with good vinegar. Each time, it is well pulverized until it resembles ink. Then one writes with this pen with a mastery of the art that is desired. Whatever coloring is desired is dyed in it. Writing is not made wide or thick; it is a plan of the middle of the pen. Then take two pots large enough for enough material for the length of a book that is to be written or a little more.[286] Both are introduced into the fire which is blown on strongly. It has previously been in contact with the sulphur of the fire. It is broken up and pounded to pieces. The two pots are removed from the fire with pincers and tongs. The material is placed in the hands and a little of good quality sulphur is added. A narrow line like the pen is made. The end of the pen is taken with the hand and suspended in this soot. It is brought close to it if there is not flame on the sulphur. If there is, then raise the pen to a small height so that it does not reach the flame. When the flame has died down, then the pen is lowered closely to the pot. The green soot is followed with the pen. That makes the effort successful for that which is desired. If it is evident that the sulphur has not burned and had not produced green soot, and the pot has become cold, then it is returned to the fire. (**63**) The other pot in the fire is then removed. Sulphur is added to the fire. The pen is then returned to the soot. This is done until the pen turns quite black. Then it is believed that it has been well dyed. Else the pot is returned to the fire with the sulphur to be heated. The white, black, and yellow places are followed by the pen without hurry. If it is achieved, it is stopped at the end. It is left a while. One may write for a short period. If the redness is dissolved while writing, then it is washed well or rubbed in a hair cloth. It is then taken out and wiped. It is examined to see whether certain places are not dyed with the black. The writing is repeated with the red on the place of the white. It is hung on the door. The work is begun as it was first described. It comes out beautifully. The correct and the complete then comes out. This is an art. Depend on the directions of the burning of the sulphur on the pot. It should not be burned on the fire else there is a flame to it with only a little soot, and it comes out light and is not useful.

Description of black writing on a white body as desired. Rely on *awīkī*[287] of which you take two parts. You also take of red lead[288] one part. It is well pul-

[282] *ṭala.*

[283] *aruz.* Probably *Oryza sativa* L. (al-Ghāfiqī, 88). Rice is a food crop thousands of years old in Mesopotamia (Levey, 46, 47). Diosc. (II: 95), ὄρυζα, speaks of its medicinal and nutritional values.

[284] *barnīyyah.*

[285] *ṭaijan maṭliyy.*

[286] This is a description of the alternate use of two pots upon which the soot is deposited. The pen is dipped directly into the desired soot on the bottom of the pots.

[287] *awīkī.* Unknown. Vocalization uncertain.

[288] *zarqūn.*

verized on a stone. Then dough of wheat is dissolved in good vinegar. It is sieved and then brought together with the red lead and *awīkī*, which (**64**) is well done to the extent the pen is rubbed with it. It is dried in the sun. When that smear is dry, one can write with the nib whatever is desired and whatever it is desired to color. Then it is suspended in the soot of sulphur as was first described. When the soot ink is attained at the end, it is thrown in water and washed well. If something remains in it, do not blacken what is desired but oil it with that well made preparation on the white of the pen and it is left in place of the black. It is then returned to the soot. This is done until the result is satisfactory.

Another description of the coloring by the pens. Red earth is well powderized, mixed, and pulverized. One can use it to write. It dries. Then it is smoked with sulphur in two good clay beakers. The writing is wiped away from the pens. What is under the writing comes out as black, the second as white.

The good pens are five in number. There is the pen for the decorative letter, the pen of the quill, the pen of the two-thirds, the pen of the half, and the pen of the third which is the lightest. They are used in the copying of lines of writing in the way ordered. Some prefer the weight of the two-thirds pens to that of the decorative letter pens although they are made similarly. The quill pen is heavier than the half pen by a sixth. It is understood that time is the essence. If the master of the decorative letter pen can write a letter in a certain time, then a master of the two-thirds pen can write it in two-thirds the time. The master of the half can write it in half the time. The master of the third can write it in a third of the time. (**65**) As to the quill pen, its time is long. However, the line is very good for these five pens. Others are inferior to it [the quill pen] such as the lightness of the two-third pen and the smallness of the one-half pen. There are also the multicoloring and the sign working, and the line in the spread dust of the *fenugreek*.[289]

TWELFTH CHAPTER

ON THE ART OF BINDING BOOKS IN LEATHER AND THE USE OF ALL ITS TOOLS UNTIL IT IS FINISHED BY THE BOOKBINDER.

As to these, there are the slab, the whetstone, the parer, the knife, and the awl, the shears, the mallet, the needles, the cutter, the press, the screw press, the rulers, and the compasses.

As to the slab, it is necessary that it be of white and black marble, the best, or some other. It is smooth on its side so that a ruler can be passed over it. It is good for scraping or binding.

As to the whetstone, it is essential that it be smooth on the surface. It should not be so soft that an iron scratches it nor so hard that it harms the iron since hardness may dull it. Some craftsmen straighten the sharpening stone, make it exact, equalize it, according to their wishes up to the handle. It remains overnight in a pot to absorb the oil which is best for it.

As to the parer, it is necessary that it be of good iron, not soft and not hard. Its measure in weight and lightness is according to the measure of the hand of the craftsman. This is also true for the leather shears. The mallet is used for the gluing process. The awl is very fine.

The shears are very straight, of the best iron, to cut leather and other things.

There are two types of needles, one that does the page sewing and one for binding the book. (**66**) The one used for sewing should be perfect and thin in body. The one for bookbinding should be shorter and thicker.

The cutter[290] should have a length of twenty[291] or less than that. The width should be good. It should not be spotted on the body and be of the best tempering. Some craftsmen do not appreciate the use of the cutter. Its handle must be adapted to the fullness of the hand. I heard that some people of this art do not use a cutter and do not beautify the art with it. They do not use it well since they have a long iron blade and cut with it in the way to which they have become accustomed.

As to the press,[292] there are two kinds. One is the press which has a cord. The Iraqis use it as well as the people of Egypt and Khorasan. The other press is the screw press. It is called by the bookbinders and by the carpenters "Solomon's binder." The Greeks call it *kḥliun*.[293] All the people of Iraq use it. As to the cord press, its length should be related to the section to be tied. If it is half-Manṣūrī[294] size, it is proper for the operation that the press be longer than the book. The book should be in the middle of the press. That is easier for the craftsman and safer for him during the pressing. The cheeks should be of good width and perfect in form. This is so that when it is desired they be closed, they hold a paper sheet firmly. The cord should be of hair newly cut when it is twisted. It should be fine and fully black, with no odor except a good one. (**67**) There should be no defect in it. It is seen in the work of the tanners in depilation. It is necessary that cord of the best hair, as has been mentioned, be used for this press. It is fine, finer than flax, and its length is enough to go around the press on all sides—four times. If one adds to this number, then it is less tiring

[289] *ghabār al-ḥulbah*. The dust was spread on parchment and then blown off. A thin film remained on which the design was drawn. It was later inked in.

[289a] Compare Appendix, chaps. 2, 3, 4.

[290] *saif*. A cutting instrument.

[291] Units not given. Probably fingers, each equivalent to approximately one inch.

[292] *mi'ṣar*. This is also the term used for the wine press and oil press.

[293] Related to κοχλιός, a screw. *Cf.* κοχλιάς, a spiral shell, helix.

[294] Bosch, G. K., in an excellent article in *Ars Orientalia* **4**: 1–13, 1961, refers to al-Jawhair who states that Manṣūrī is an Egyptian paper of large size having polished surfaces.

to the craftsman because as often as one adds strands the less difficult it is to pull. When the cord is in two strands on every side, it is necessary that the stick be twisted many times. But if there are four strands, it is turned less than eight times. When there are more than four strands, it is twisted less than that or four turnings. The length of the stick is according to the length of the finger in that it is light, thin, and smooth. It is necessary for this press that the two cheeks have slots in the places where the cord is to be. It is better when the cheeks of the press be shortened. This is so that when the cutter falls along the edge of the press it does not cut off a bit of the cord.

The best straightedge[295] is made of ebony or boxwood.[296] For drawing, inking, and lining, it is better that they be of these materials. As to the straightedge for the usual work, it is desirable that it be of the wood of the willow. The willow should be on the edge, that is, on the sides of the straightedge. This is to avoid damage which could happen to ebony; it would affect the lines by the imperfections of the straightedge. (**68**) As to the straightedge for drawing, it is necessary for it to be very long. It is not made thick or thin. Straightedges for inking are very thin because they are worked by two fingers. As to the straightedge for lining, it is necessary that it be like the latter in thinness and lightness. Lining will be noted in the chapter on pretty drawing. As to the folder which is the "straightedge of the air," it is the one used for work on leather and in its craft to force the air out from under the leather, to correct wrinkling and crookedness and to straighten the leather on a level surface. It must be of good thickness and a span in length. It is of wood of the best oak. It is square and thin at the edges so that it smooths the leather when it is passed over it. The handle is made of oak since ebony and boxwood, if pounded on the press, have their edges dulled and they break.

For the divider to be good, it is necessary that the body be light. The divider marker[297] should be thin so that it makes a fine line and the opening and closing joint of the divider should be accurate. If it is not correct, it is necessary that it be adjusted. The divider is to make the suns. These are circles in pretty drawings that are in the middle of the book. The description of it and the description of the work will be mentioned.

Then there are the irons for tooling. These are the gouge, and the "breast" which is called "the breast of the falcon." There are also the ornament, the dot, the "encircled," and the polisher which is called a *dast*. Then there is a fine polisher. There are different stamps. The dots for impressing will be mentioned in their place. This is the total of instruments. It is complete. (**69**)

One who seeks this art should have quick understanding, good observation, dexterity of the hand, and be certain without being hasty. The latter is a good manner of getting along and it has the elegance of attracting others of grace and good character.

The first thing to do to begin this art is to place the part to be sewed beside you on a slab. It is put to your left. A quire is picked up with the left hand. It is opened with the right hand. It is put down on the slab and opened. Then the folder is passed over its center where the binding thread is to be. Then it is folded and the end paper is cut properly. This is a double sheet; one page is pasted on the leather and the other remains on the quires to protect the book from harm and dirt. Then this is done to the remainder of the quires until the last. When finished, the thread is then twisted for tying. It is in three strands according to the measure of the fineness of the thread. It is best if the thread is fine for then the twist is best. If it is coarse, this part is damaged since it turns in every sheet and produces extra bulk. If it is coarse and the book is bound, the press will fall on the end of the thread which remains and will leave a mark. It is similarly so if one winds a thread on his finger to the end. Thus it causes extra thickness in the interior of the book.

There are ways of bundling (gathering in sewing). Some are used by craftsmen for ease and quickness. (**70**) It is that the needle penetrates two places. Others work with two needles or three. I saw the Greeks do it but I cannot approve of it. I cannot describe it. When the section is tied together with string, the place where it is bound is then pounded with the folder previously described. It is then put between the knees. The press is taken and one of the cheeks is put against the left knee and the other against the right. The book is in the middle between the knees. The end of the cord is put in the left hand and wrapped around the press. Both ends of the cord are tied together. Then it (the book) is taken from between the knees while it is in the press. The protruding spine is put on the slab. Then the ends of the sections are pounded with the folder where sewn until even. There is no difference between where it is and where it is not sewn. Then it is raised from the knees. The two sticks, called the *marāwīn* are tied; these are tied lightly, but not too much, since, if the tying is too strong, the spine of the book is twisted and harmed. The paste is then dissolved loosely. A small pot is taken and a little water poured into it. In it is thrown some paste. The paste[298] is beaten and stirred with the middle finger of the right hand. It is loose, not

[295] *misṭar.*

[296] *baqs*. In Maim. (9) *baqs* is equated with *simsār* or *shimsār* (I. B. 315, 1342). There seems to be uncertainty in application of the words *simsār* and *shimshār*. *Baqs* was used by the Arabs in applications for contusion of the head, and headache. The box is given by *DAB* (348 ff.) as *ŠIM.ŠAL* in Sum. and *shimshalū*, *shimeshshalū* in Akk. It was used in Babylonia as a stomachic, for the head, and in enemas.

[297] *baikar*. A compass divider.

[298] Al-Muqaddasī states that in South Arabia, as in Egypt, the quires were glued together and the volumes cased with wheat starch. In Palestine the asphodel paste (*ashrās*) was used. In Yemen only wheat starch paste (*nashā'*) was used. de Goeje, *Bib. geog. arab.* **3**: 100.

thick. It must be fast in drying. Then a leaf is taken, folded, and cut in the middle. Each half will be the width of the spine or more than it by two fingers. (**71**) The glue is taken up with the middle finger, the rest of the fingers hanging, and the spine is smeared lightly, letting the slime of the glue fall on the book. There does not fall inward a bit of it from the spine. Then the sheet is applied so that the excess width is on one side. The spine is smeared again. Then the other half sheet is pasted in the opposite way. The reason I say the opposite is because the excess falls on the other side. Then a sheet is placed over the spine. It is grasped with the left hand and smoothed because, if the folder is put on the moistened sheet, it will be pulled and spoiled. This is from the mysteries of this learning. When this is done it is left in the air or, if desired, in the sun. If there is any haste, it is put next to a low fire. It is not pulled until it is dried out evenly or else it will twist. Be careful. It is necessary that a measure has been taken of the book before it is left in the press. The pot is put beside you on the slab and the paste is applied as desired. Then the sheet is covered by another. It is left over it. It is rubbed with a cloth and then the folder. Then more (paste) is cooked as it becomes necessary. The Iraqis paste the book cover to the pages without the end papers. It is called the *tāqawi*. Others who have seen this believe that it protects the book and it is like cloth and board in strength. When the book is dried, the board is strong. It is taken out from the press with care. (**72**) It is put on the slab and over it is folded the excess of the two sheets (hinges). Then the boards are well smoothed after which the straightedge is put on its edge and a line drawn along which it will be pasted. It is stuck together in part, that is, by removing the sheet that was pasted at the lower part and then putting the board on the book. Then it is pasted. When it is pasted, it is done on the two sides. A long narrow sheet whose width is two fingers is pasted on it from the other side to prevent it from being opened excessively. When this stage has been reached, the leather is applied to it.

As to the leather, it should be unspotted, and, if it is imported, Yemenite. It is made differently from the Ṭā'if and like the one made in this district. It is essential that it be clear and pretty of color. It is nicely tanned so that if it is rubbed in the hand and appears soft, then it is the best. If it is not like that, then it is not good. It should be washed in a warm bath for warmth opens the pores and makes it soft. In the case of imports from Ṭā'if province, the water should be salty since they tan with salt water. If it is washed with warm water, its oil comes out to enhance its beauty. As for the leather which is Egyptian tanned by the method of the Yemenites using gallnuts, it is washed with sweet water because it is tanned with it. If the leather is for tooling, then it is flexible, light of weight, and is less than two pounds. It is of good tanning. If it is to be plain, its weight is two pounds and it is grained on the (**73**) surface. If it is of this description, then it is washed in a clean place. It is necessary to use caution so that anything that blackens it like iron or a nail will not be in contact with it. When the gallnut leather is washed, it is rubbed well on the surface with a potsherd to get rid of what remains of the gallnut and the gnawed. It is pressed out well and put face to face. Then it is opened out until the above has disappeared. Its ends are cut off. It is then cut according to the measure desired. It is spread out on the slab and rubbed with the straightedge previously mentioned. If anything separates, then the outer surface of it is scraped off. The best scraping[299] is done when it is nearly dry. Then the knife does not pull it as it does when it is dry. When it is scraped, it is necessary that there should not be any scrapings under the leather. It should be cut on that place. When the scraping is finished, then comes the washing. It is washed until the water comes out very clear. If it is seen that the water stays in spots on the surface of the leather, know then that there is an excess of fat. It will not make a good job. If it is desired to remove the oil from it, two ounces of powdered gallnut are thrown on every layer. The piece is stretched between the hands. The gallnut is spread and all of it is moistened over and over again. The leather is returned to a pot with water in it and it is covered with more water. It is made heavy by something so that it cannot float, and it is left in a night or a day until late at night. It is then removed from the water and rubbed well. If this can be done with a bit of bran, it is better.

(**74**) If the leather lacks proper tanning, it blackens and is oily to the touch. Carry on as has been done with the fat. It is good for it.

As to the nature of the gallnut in the leather which is supple, it hardens it; if it is hard, it softens it. If there is fat, it removes it; if it is without fat, it provides some. Understand that. Then dye it.

Description of dyeing leather and paper. The red in dyeing is of many types. In one, the best possible sapanwood is taken. There are two types; one is the "little" and the other is the "princely." An ounce is taken of the powder. It is immersed in water a night or a day. It is then put into a copper pot, a clean utensil. On it is poured ten *raṭls* of water and the best powdered, sieved wild *qalī*. It is then boiled on a good fire until half of the water is lost. The essential of the process is that a rod is left in it. Drip it on your thumb. If it remains and does not drip, then it is successful. It is taken down and purified. If desired, this may be repeated on that type which is sold. The first is the better of the two. It is left until it cools. Then dye with it. For dyeing, paper is put in the solution with care and then spread out in the shade. For leather, put sapanwood in a vessel to which has been added water of sapanwood. It soaks it up. A hair brush is put in the water of sapanwood, or a piece of felt is wound on

[299] A paring operation.

the head of a rod and immersed in the sapanwood, then rubbed on the leather. This is done twice or thrice. Then the leather is pressed and the dyeing is repeated. (**75**) Wool is dipped in the alum. It is necessary that the alum be moistened immediately before the dyeing. There are varieties of alum; the best is that which is sour when tasted by the tongue. If it is salty, then it is not good. It is then soaked in whatever is desired; if it is sharp a little water should be added until it is corrected. Then the wool of the brush is immersed in it and it is passed through the sapanwood. It is rubbed well and left, soaked and then spread. This is repeated until the depth of desired redness is attained. It is then spread on the slab and the end of the straightedge is passed over it. If one is at hand, a piece of cloth, a hard woolen rag, a haircloth, or some other thing is used. It is suspended until it is dry. If desired, it can be dyed black. It does not beautify it. It is wet-dyed in the process of dyeing it black.

Description of dyeing black. A vessel glazed both inside and outside with the best glaze is used to hold nail heads cleaned of rust. It is filled with vinegar and left two or three days until it has reacted. It is better if pomegranate rind is used. When this is done, then a small stick with a piece of wool or felt wrapped tightly about it is dipped into it. The dye is applied with it. Take care not to touch the hand or it will turn black. If it does, then to remove it, dip the hand into lemon[300] juice which also removes sapanwood. (**76**) It is repeated once or twice. It is then rubbed, washed without delay else it may be burned and thus wasted. When washed, it is then scraped and returned to the washing. The dye goes on that which was traced.

If it is desired to make a prettier black, then it is necessary to use yellow myrobalan[301] juice or pomegranate juice which was obtained by dipping it in water until its color has come out. It is immersed in the solution while it is wet. It is left until it is dry.

If it is desired to dye it yellow, then there are two of these colors, orange and yellow. As to the orange, the *ᶜakkar* goes with saffron. Leather is dyed with it. The leather, however, should be all wet or completely dry so that it is not spotted. If it is desired to dye with *ᶜakkar* alone, then the result is different. If it is saffron alone, then it is yellow. This color, all of it, is immersed in water of yellow myrobalan. When it is soaked, the leather is softened. The hair brush is used to go over it if it is to be multicolored. If it is Indian spikenard, then the fibrous part is prepared in two ways. One way, of Raqqah,[301a] is to clear the color of the hair. Another type of this is the Antiochan which has thick hair and is dark brown in color. If it is to be green, then it is dyed with *ḥirāq*. The *ḥirāq*[302] is a flower of a species of the cucumber. It is a greenish blossom. A twig of cedar is rubbed with it. Then it is suspended on baskets under which old urine has been left. If it is desired to dye with it, the twig is put into the solution which will come out a beautiful blue. It is tested with the finger. If it is pale, add *ḥirāq*. If it is deep, then water is added. It is used to dye as the yellow was used. It comes out a beautiful blue. (**77**)

Description of dyeing with *ᶜakkar*. As to the *ᶜakkar*, in the beginning of the procedure, the best Carthamus, which is soft and penetrating, is dried and pulverized in the mortar, and sieved with a hair sieve. It is then put in a vessel and water poured over it. The hand is left in it and it is stirred well. Then a woolen handkerchief is put on a wooden frame and the Carthamus is poured onto it until the water flows out. It is returned. The hand is introduced into it after the water is poured in. That water comes out from under it. It is thrown away. Then water is poured on it and the hand immersed in it until the hand is left clean. Then the handcloth is removed and tied up firmly and left on a stone. Another stone is laid above it, or a heavier stone which is sat on until all the water in it has run out and a dry residue remains. The handcloth is untied. Sit and stretch your legs. Then your left hand is put on the handcloth. A little Carthamus is taken, then the hand is opened. It is so done with all of it until no more is left. That which was opened is returned. It comes out this way. Then thirteen dirhams of the best wild *qalī* are pulverized and ready beside you. It is thrown on five dirhams of Carthamus and it is all mixed with the hand. Then the same thing is done with five more parts. It is all rubbed until the dye comes out in the hand. If the hands have been reddened by it, then know that it has reached its limit with the *qalī*. (**78**)

[300] *līmūn*. The lemon has not as yet been identified in Babylonia. In Arabic times, three parts of it were used—the rind, pulp, and seed. It had various uses in the Arabic materia medica according to ibn Jamīᶜ. No mention is made of any other part of the plant being used. The juice mentioned is probably from the pulp. The lemon is *Citrus limonia* Osbeck.

[301] *halīlaj*. There are four kinds of myrobalan, yellow, Indian (black and small), Chebulic (black and large), and the Chinese (I. B., 2261). *Halīlaj* or *ihlīlaj* comes from the Persian *halīlah* which may come from the Sanskrit *harītakī* (al-Ghāfiqī, 264). The yellow kind is probably *Terminalia citrina* Roxb. (hara nut). It is a stage in the growth of the Chebulic myrobalans as are the other myrobalans. The belleric myrobalan (*T. bellerica* Roxb.) is *balīlaj* (al-Ghāfiqī 123, *Tuḥfat al-aḥbāb*, 43, 122, 126). Emblic myrobalan, *amlaj*, is treated by al-Ghāfiqī (13). These were unknown to the Greeks but were known early in India (Ainslie **1**: 236–241).

The "three fruits" of the myrobalan are distinguished as follows: when very immature, it is called Indian, when half mature and still yellowish, Chinese, and when yellow and quite mature, Chebulic. The unripe fruits contain 20–30 per cent gallic and tannic acids and a greenish oleo-resin, myrobalanin. The myrobalanins are still sold (Ducros, 13, 14, 15) as intestinal astringents and purgatives as well as tanning agents. According to Ducros, the Chebulic and Indian types come from *Terminalia Chebula* Retz. while the yellow is from *T. citrina* Roxb. Al-Kindī knew *amlaj*, the emblic myrobalan, in musk recipes. Meyerhof (Maim., 81) says that this is the fruit of *Phyllanthus Emblica* L., a Euphorbiacea which has no relation to the Terminaliae.

[301a] A famous city of Arabia. *cf.* Yāqūt.

[302] *ḥirāq* (vocalization ?). It is the green flower of the *maqāti al-faqqūs*. The *faqqūs* (I. B., 1690) is the cucumber.

Otherwise, it is repeated until the hand is dyed and becomes red. It is returned to the carrier and water is run on it in an amount to cover it. A pot is left under it so that its water flows into it.[303] The water that flows out is taken to something else. Every time the water diminishes above, water is added until the water coming down appears clear. Then it is removed. If desired, it is made with wine vinegar. Two ounces of the best wine vinegar are thrown on and stirred with a rod. Water is sprinkled on it with the hand. It is immersed in water a night until it settles. In the morning, the water in it is decanted and used. If it is desired to use water of pomegranate, then four ounces of pomegranate seed is put in four *raṭls* of water and left immersed for an hour. It is decanted and the same amount of vinegar is added. If the *ᶜakkar* is handy, the water is decanted and more poured on it. That will keep it.

To return to the description of the drawing. When the leather is dry, it is necessary that the book be trimmed equally with the cutter which is called the "trimmer." The book is put between the hands to do this. Only some craftsmen do as I am describing. A straightedge is put diagonally on the book. Its middle is marked. Then the straightedge is turned back to the other sides and the same is done. Thus, in the middle of the book, a cross is formed. One leg of the divider is put down on the intersection of the cross. Its other leg is opened to the side of the book. This is the description of the binding. Nothing was left unexplained and (**79**) unmentioned of the instruments for binding.

Description of solution of glue from leather scraps[304] from whatever animal it is. Its hair is shaved and then the material is soaked in a vat or large pot. It is covered with two spans of water. It is soaked until disintegrated. Then it is left on the fire until it grows cold. It is clarified with a woolen cloth sieve. A cover is put on until it is cold. It is cut into small pieces with the knife. It is strung up in the sun. It comes out as very effective.

Iron ends like the heads of nails or similar, are put in water of myrtle, or water of the rind of the pomegranate or the macerated gallnut in water of myrtle or rind of walnuts. Any one is suitable, all of it or separately. It is left in the sun and stirred with a palm branch at times. Without vitriol that water turns black. If it is desired, it is thickened. From it there comes black dust. It takes the place of soot. When desired, it is dissolved in order to work with it. If it is desired, plant sugar[305] is added to the gum to thicken it. It is beautiful. If it is desired to make it smell, good clean frankincense[306] and saffron are added to thicken it. A thickness remains which does not dissolve alone in humidity like Egyptian soot ink. When it is desired to write with it, it is powderized and put in the *līq* which may be recalled. It is soaked with clear water of myrtle.

In all these procedures, the flowing water is allowed to settle until it is clear to the extreme. It is necessary that there be near you a water vessel—its mouth like the mouth of the bugle—to make good the pouring into it. (**80**) There is always soaked with water of clear myrtle an inkwell full of black ink. The odor of the *līq* becomes good. The color of the ink becomes pretty, and strength is added to it by addition. If myrtle is cooked in water in which sorrel has been cooked, then it is better. The dry myrtle and the green are the same in this process. If desired, it is pulverized, and, if it is desired, its leaves are left in their state. As to the rind of the pomegranate, it is not introduced in this work except when dry and as much as previously. It is better if it is older. The red anemones are also introduced in this work, i.e. the red of the blossom. The black is cut from it and thrown away. Only the red is used. In this work are also introduced the green St. John's bread and leaf of the tamarisk.[307] Any of these may be used to take the place of gallnuts. If it is all gathered together, it is stronger. The effect of its print on cloth comes out only with trouble. In the manner mentioned, things made from it are good since it enters into the work of collyrium.

Description of a compounded ink. The green myrtle is boiled, dried, and purified. Then one part of gallnut, one part of gum, and one-fourth part of vitriol are each brought to an extreme fineness. They are brought together and on them water of myrtle is poured. It is rubbed to the extreme. If they are all equal, for each dirham of gallnut, a dirham of the best soot is taken or

---

[303] This is a filtering process through the pores of an earthenware jar. Here is described a repeated washing and filtering process.

[304] In a Sumerian tablet of *ca.* 2000 B.C., glue (*ŠE.GIN*) for leather decoration is made from leftover leather pieces (*ZAG.BAR*) probably by a similar process. This glue was applied to leather to make it adhere to furniture, doors, and chariots. *ŠE.GIN* was known in a dry or liquid form. Dyes were frequently added to the glue to be applied decoratively to leather (Levey, 77).

[305] *sukr al-nabāt.* From anyone of a number of plants which yield sugar in quantity.

[306] *kundur.* Diosc. (I: 68) states that frankincense, gum of *Boswellia Carterii* Birdw., λίβανος, is grown in Arabia. It had many uses medicinally from Egyptian and Babylonian times onward. Al-Kindī knew *kundur* in his perfumery recipes under the name of *lubān* (62). I. B. (1974, 2012) states that there are many forms of incense useful in many directions. Jābir also discussed *kundur* (64*b*). The *Tuḥfat al-aḥbāb* (214) states that the best is from India and Syria.

[307] *athl.* In Sum. [giš]*ŠINIG*; in Akk. *bīnu.* Probably the *Tamarix orientalis* Forsk. = *T. articulata* Vohl. The word in Syriac is *bīnā.* In Babylonia the wood was used for dishes, to make eating tools, and in medicine for the eyes. The seed was used as a stomachic. The "water," gum, and root were also used medicinally (*cf. DAB*, 272 ff.). The tamarisk leaves were used for washing (Levey, 123). Diosc. (I: 89) mentions tamarisk, ἀκακαλλίς, as an Egyptian tree whose infusion is good for salves for the eyesight (ibn Ghāfiqī, 69). Maim. (9) states that the "*athl* is a species of tamarisk (*ṭarfā*) whose seeds (*ḥabb al-athl*) are what the Egyptians call *al-ᶜadhba*, "the savory." The *Tuḥfat al-aḥbāb* (23) says that *athl* = *tākkawt.* This may be incorrect since the latter indicates rather the gall of the tamarisk. Two species of *Tamarix* are today grown in Iran and Iraq. Both of these are used in various ways (Hooper, 175, 176). *Cf.* also Ducros (56).

some other. It is brought together with two or three dirhams of flowing honey or flowing molasses.[308] If the soot disappears and unites with the honey, more is added and rubbed well, then purified, and removed for the (**81**) time being. It is shaken as often as a bit of it is taken. If it is left in a plate or something like it for a night, then a film comes out on it. It is necessary to mix it again. Ibn Ghaṣīn said that the gum dissolved with water and the sweet is not necessary. He said, "The sweet spoils the ink and moistens it." It is said that in the case of the rind of the pomegranate, a man made it to sell it. Books, which were glued with it in it, many of them, were ruined by the glue and also by the strength of the honey. He said, "Like that is the procedure with the rind of the gallnut. As to the things that are introduced into ink and do not glue such as the gallnut, myrtle, tamarisk, yellow myrobalan, and red part of the anemone of which the black is thrown away, these are all found in the compounded ink." It is said that if a small part of the pomegranate rind is used, alone or added to, only the inexpensive of the mixture need be used with it. That is why it glues. When all of these mixtures are present together, it is better than each alone.

Description of a compounded soot ink. Dry myrtle is put into water in which sorrel root and leaf have been cooked. It is boiled, filtered, settled, and in it is soaked dry pulverized rind of pomegranate. It is heated once and cooled once so that it is not made too viscous in the water. If it becomes viscous, water of myrtle is added, then filtered, and allowed to settle. Then powdered gallnut is added. It is heated once down to the warmth of the sun. It is filtered and allowed to settle. Powdered vitriol is added to it and it is shaken for a day. (**82**) Then it settles from the morning on until utmost purity is achieved. Pulverized gum, not powdered, is added and also Egyptian ink—both are dissolved in some of it. When they dissolve thickly, then all is mixed and raised into a glass vessel which is open at the mouth for air. It is shaken as days pass; then these waters are left to settle. To aid in this process, a woolen or felt cloth is used for clarification. As the mentioned mixtures were made in water of myrtle, mixing after mixing, the same is done with the remainder of the mentioned mixtures introduced into this work until it is all gathered and again becomes a clear water. If the water becomes viscous in the mixing, then water of myrtle is added until it flows, settles, and becomes clear. The wisdom of this work is in the clarification and settling. If a residue remains in the filter or on the felt, then clear water of myrtle is added. It is shaken and settled until all the strength is removed from the residue. If the waters collected are clear, it is made into a syrup in the sun or on a fire like the heat of the sun until it assumes normal viscosity for writing. If this method is followed, then no vitriol is added until all is put together except the Egyptian ink dissolved in water of myrtle, both of them clear to the utmost. It is all then made into a syrup which has strength for writing. It may be desirable to put in frankincense powdered to the extreme, and a little white sugar. If it is rock sugar, then it is better. It is put into a glass vessel[309] having a wide mouth. (**83**) Its mouth is like the mouth of the bugle, not on the form of the water vessel which is made up of two glass sections. It is stirred every day. If some of it is taken to write with, then it is stirred alone and shaken until no sediment is present so that its thickness and thinness may not ever appear in the writing. This is the reason for it. There is always present pounded not rubbed gum which is thrown into the inkwell as often as the brightness diminishes. The *līq* of the inkwell is stirred for some time so that there is no sediment in the inkwell. One does not write except with *līq*. The best of it is that felt which is between soft and hard. A little circular piece is cut off and it is then brought together with a stitch in the middle. This should be done in the middle three times. The *līq* becomes between soft and hard. It is turned in the inkwell as often as one writes. It is then safe from sediment and residue. When something syrupy gathers around the edge of the inkwell, it is scraped with a knife. It is thrown into a vessel of the ink and nothing is lost. The gum is to enhance the lustre; the gum accompanies the pen in writing. If the gum is excessive, it is necessary to add water and it is made as mentioned. This method that I have chosen is a general one followed for all inks and dyeings. This book is finished with thanks to the generous Allah and his overwhelming assistance. May Allah's prayers and peace come to our master Mohammed and on his family and companions. Thanks to Allah, the master of all the worlds.

Note[310] on *lāzward*. (**84**) Generally, a little is probed on white cloth. It is rubbed and the dust shaken out. If the cloth is dyed, then it is false.[311] Or a little may be put in water, rubbed, and left an hour. If the water is dyed, then it is false. Or a little of it may be kneaded in the hand with saliva. It is left until it is dry. It is shaken out. If it dyes its place, then it is false. If its place remains as the color of the hand, then it is genuine. Or it is formed and put on a copper sheet or on the back of glowing charcoal for an hour. If it burns or becomes black, then it is false. If it remains in its condition, then it is good. As to the test in regard to heaviness and lightness, the light one is false and the heavy is better. But this latter can be falsified with some stone. It doesn't show except through the fire.

Note on the testing of verdigris. The types are Iraqi, Emesan, Egyptian, and Greek. All are verdigris of

[308] *dibs sāyil.*

[309] *faqqaʿah zajāj.* A glass vessel having a very wide mouth.

[310] Here follow a number of notes relating to the text. Some, however, seem to be extraneous.

[311] I.e. ersatz. It was not uncommon for the ancients, as it is today, to falsify expensive ingredients.

copper with vinegar or with vitriol. The best is that obtained from the sprinkling of a copper sheet and then kneaded into a mass. Something is mixed with it to stand with it. Then it should not be used in drugs for the eyes. The one which is good is light of weight, quick to break up as glass is broken up. There are in it white crevices. The Emesan type is inferior to the Iraqi, and the Egyptian is inferior to it, and the Greek is inferior to all.

Note on testing white lead.[312] Some is Greek and some North African. Essentially it is the flower of the lead which deteriorates by means of vinegar. The best of it is strong in whiteness; it does not tend toward blue. If some is rubbed between the fingers, it is found soft and heavy of weight. The falsified is the opposite of that.

(**85**) Note on testing good mercury. This is the kind that is always moving and white. If it is stirred with the finger, it will not separate. If the hand is put in, it will not affect it. It is free of odor. For the false, the reverse is true.

Note on testing rosewater. If it is desired to know whether the Nisibus rose solution is good or false, sweet water is poured on a little bit of it. If it whitens like milk, then it is good. If otherwise then it is false.

Note on testing good opium.[313] A bit of it is taken into solution with water and then filtered. If a residue remains, it is false; if otherwise, then it is pure. The odor of the pure substance is strong. It is white tending toward red. In its taste it is somewhat bitter and distasteful. The false is the opposite of that.

Note on testing musk. There are many types of musk.[314] Those known are five: Indian, Bahārī, Tibetan, Iraqi, and musk of the *bad*[315] plant. The Indian is black of color tending toward a little red. The poor part of it is black without the red. The false part of it is that which tends to redness. If it stays for a long time, it heats up and worms develop in it. Its falsification is made to appear when it is pulverized in rose water and it is left until it is settled. If there is a sediment, and if the water is white or cloudy, it is true. If the water is poured out and there is no sediment in it, then it is false. As to the Iraqi, it is blond in color. It is pulverized and put into a container which is closed. If its odor is unusually strong and if there is in it no taste, it is good. If it has a different taste, it is something else to which the taste belongs.

Note on testing the Tibetan *tanbitī*.[316] All musks have the attribute that they can be made fine by rosewater or by water (**86**) except the pure *tanbitī*. It can be pulverized by pounding but not by rubbing. It is hard, heavy, and fatty of odor.

As to the Bahāri, the *tanbitī* can be falsified with it. The difference between them is that the *tanbitī* is black and its crack is black; the Bahārī is lighter and its crack is white. It is inferior to the Iraqi. As to the musk of the plant, it is fathered from the *bad* of the Indian country. It is imported. There is no test for it for it is the most inferior of the musks.

Note on testing butter. A bit of it is taken on a rod and then brought near the fire. If it flows, then it is false. If it is gathered up and is contracted, it is good. It can be falsified with something. If the fire smells of the known odor and there is in it an unusual odor, then it has been falsified with that unusual thing. A bit can be rubbed on the hand until it is warm and it smells. If there is in it something false, the odor of that will appear.

Note on the testing of crude ambergris.[317] It is left on the fire; if it boils, then it is falsified. If it does not boil then a residue is found in the lower part of the pot. Its taste is determined; if it is salty, then it is falsified. As far as the feeling is concerned, the light one is pure and the heavy is falsified and is oily. It has in it wide holes. The fish swallows it, then throws it back. Thus it acquires a fatty smell and heaviness. As to the

[312] *zahrat al-raṣāṣ*.

[313] *ufiyūn*. In Babylonia there were many uses for the poppy or opium. In Sum. it is *ŠA.LAM.BI.TUR.RA* and in Akk. ^{šam}*araru* and other synonyms among which are ^{šam}*PA.PA.PA* = "tops" of the poppy. Opium was used for the eyes, as a stomachic, and in suppositories and enemas. All parts of the plant were used for various purposes (*DAB*, 227 ff.). *Ufiyūn* probably comes from the *Papaver somniferum* L. still grown in Iran and Iraq (Hooper, 147–148). Diosc. (IV: 64) discusses opium poppy, μήκων ἥμερος, as an important simple for the stomach and menses. Compounded with other simples, it is used almost over the pharmacological spectrum. In Arabic times, Maim. (35) listed it as being called *al-marqad*, the soporific. I. B. (116, 2120) gives many medical uses. Ducros (8, 98) states that opium is still considered as a universal panacea. It is usually eaten, not smoked. It is taken as a paste or sweet known as ᶜ*maᶜjūn* or *manzūl*.

[314] *misk*. It may have been unknown in Babylonia. In India, on the contrary, it had its origin. According to Māsawaih (398–399), musk is one of the principle aromatics; it is good for the heart. Al-Kindī (72) employed musk in his perfumery recipes and drugs frequently. Musk is mentioned in the *Talmud Babli.*, *Barochat* 45; *Yerushalmi*, *Barochat*, Per. 6. Ibn Māsawaih (d. 877 in Samarra), in his treatise on Simple Aromatic Substances, mentions musk as one of the major aromatics. Much of his work was copied later by al-Nuwairī. See Wiedemann **49**: 26–30. *Misk* comes from a similar word in Persian. It is a secretion found in a vesicule at the prepuce of the male Moschus maschiferus common in lower Tibet. Masᶜūdī, in his *Murūj al-dhahab*, (**1**: 353) states that the Tibetan type is better than the Chinese. Ainslie (**1**: 228–230) states that in India musk has long been used as a stimulant and antispasmodic. It is also used in dyspepsia and typhus. I. B. (2127) states that musk is used also as an antidote to *al-ḥalḥāl* and *qarūn al-sunbul*. It has an opposite effect to that of camphor. For musk compounded. *cf. Tuḥfat al-aḥbāb* (280–379).

[315] Unknown plant. It is perhaps *badah* or *badad* (al-Ghāfiqī, 178). *Badad* is listed in I. B. (253) as uncertain.

[316] Vocalization uncertain.

[317] ᶜ*anbar*. To the Arabs, its pharmacological value lay mainly in its odoriferous property when added to other drugs. Māsawaih (400) thought ambergris to be one of the five principle aromatics. When used in electuaries Māsawaih states that it is a remedy for humours of the aged. *Cf.* Wiedemann (**49**: 30–32, 56, 330–332) and Masᶜūdī (I: 333 ff.) for opinion of the Arabs. It was still used as an aphrodisiac in Bengal. (*cf.* Ainslie **1**: 15–17) in the nineteenth century.

kneaded part, if what was pulverized is powdery, then it is falsified; if it is pulverized and has elasticity, it is good. It can be falsified with something which has elasticity. A bit of it is put on the fire. If its odor is allowed to rise, and then if its powder is clear and soft like ash, (**87**) then it is genuine. If it contracts like hair when it is burned, and in its feeling there is a little hardness, then it is falsified.

Note on testing aloeswood. The best of it is black, heavy, and plate-like. When it is left on a glowing charcoal fire, it comes out oily. The false is the opposite of that, Aloeswood comes in various types, *qāqal*, *nakī*, *raṭb*, *talūn*, *nī*, and *ṣanfī*. The best is the *qāqal* and the *raṭb*. It is taken while green and put in honey to keep its strength. It remains green. As to the *nī*, it is steamed. I have not seen it until now. The *ṣanfī* is the closest to the *qāqal*. The *nakī* is inferior to it. For that the *danū* is known as the *huwaidī*.

Note on testing of theriac.[318] A bit of it is taken and put on coagulated blood. It dissolves it. If it does not, a drop of milk makes it thick.

Note on the testing of the good *Qaṣūrī*[319] camphor. Its color is white tending toward this extreme. In the earth, it is just about black. It is falsified with *riyāḥī* and *nāza*. To distinguish them, the pure *nāza* and the *Qaṣūrī* can be rubbed while the *riyāḥī* cannot. It can, however, be kneaded; the odor is bad. *Qaṣūrī* and *nāza* have a slight smell which is not bad; they do not have the ill nature. The *Qaṣūrī* is easily rubbed but is not diminished when rubbed. The *riyāḥī* can also be rubbed but the rubbing is coarse and the color is white. When it has been in storage for a few days, a green is acquired with which the *Qaṣūrī* is falsified. It is distinguished by its heaviness and coarseness in rubbing while the *Qaṣūrī* type is light and soft in rubbing.

Note on testing oil of balsam.[320] A small amount is dripped into water, drop by drop. (**88**) If it spreads like other fats, then it is falsified. If it remains firm in its place, then it is genuine. Some of it may be put in a cloth. If it remains fixed in its place in the manner it was placed, then it is good. If it spreads, it is falsified. Or if it is touched with a rod and it is lit in the fire, then it burns and has a good odor.

Note on testing almond[321] oil. It is soaked with hot bread. If its taste is good like the taste of the almond, then it is genuine; if otherwise, it is falsified. It may also be rubbed on the hand, its odor smelled, and its taste investigated. If it does not resemble the taste of the almond, then it is falsified. The color of it is that of almond oil, a yellow tending a little toward white. The false type is yellow tending toward a little redness. Know that.

Note on the testing of the buckthorn.[322] The types of buckthorn are Indian, *ṣanafāwī*, Meccan, and another kind called *zabal*. The Indian may be falsified with the *ṣanafāwī* since it is close to it in properties. The distinction between them is that the Indian type shines in a broken spot and has a light touch. It makes a yellow mark on the probing stone. It has a bitter taste. It can be falsified with a little *mubarus*. One can distinguish between them. The bitterness of the *khūlān* is a little annoying; the bitterness is at the first taste. The falsified, however, is strongly bitter without being astringent, and is heavy without being glittering. The *ṣanafāwī* can be falsified with the good Meccan type which is known as *fatr*. They are distinguished with the use of the probing stone. If it is *ṣanafāwī*, then it is less yellow than the Indian. The Meccan tests green with a little yellow. In its taste, it is salty.

(**89**) Note on the testing of the clove,[323] walnut, myrobalan, and the costus.[324] These are each made fine,

[318] *tiryāq*. From the Greek, θηριακή. The Persian is *diryāq* (Lane). Jābir (VI: 171*a*, 176*b*, 177*b*, 188*b*) discusses a theriac of four drugs and also one made of five which can be used as an antidote for poisons and snake bites (VI: 190*a*; VI: 191*a*).

[319] Many types of camphor are described by authors quoted by I. B. (1868). *Riāh* was the name of a king who supposedly discovered the camphor named for him. *Qaṣūrī* is the name of a country. *Nāza* is not mentioned in I. B.

[320] *balasān*. Various balsams are known in India (*cf.* Ainslie (**1**: 26–28). Jābir used it in his book on poisons (VI: 171*a*, 189*b*). It may be the *Balsamodendron gileadense* Kunth, at one time cultivated in Arabia, northern Ethiopia, and Egypt. It is the *βάλσαμον* of Diosc. (I: 19) used for ointments, for asthma, pleuritic difficulties, sciatica, epilepsy, and other ailments. I. B. (336) devotes a long section to the many uses of *balasān* for digestion, scorpion bite, facial tic, and other diverse troubles. Maim. (324) confused *balasān* with *qulqul*.

[321] *lauz*. The bitter almond is probably ${}^{i\check{s}}$*sīrdu* in Akk. and *šār'dhā* in Syr. However, *lauz* is more likely the sweet almond ${}^{\check{s}am}$*šiqdu* = ${}^{\check{s}am}$*nusḫu*. The former is the Akk. cognate for the Hebrew *šāqedh*. *Nusḫu* is related to the Ar. *lauz* (*DAB*, 254). The sweet almond had many medicinal uses particularly in confections and oils. The Hindus, it seems, did not use almonds as medicines (Ainslie **1**: 6–8). Al-Kindī (61) used *lauz* together with jasmine oil, costus, and sandalwood to make a salve. Garbers probably has confused the bitter with the sweet almonds in his discussion. In I. B. (2040), Leclerc does not give the genus and species. In mediaeval Arabic times, almond, particularly the oil, was used for the stomach, to increase the sperm, for the intestines, and for all dryness. There is a large almond crop in northwestern Persia. It is *Amygdalus communis* L.

[322] *khūlān*. Lycium. For three types, *cf.* I. B. (680) with Galen's statement. One is Indian, another is Arabic, and the third is prepared from wood of the *zirshik*, barberry thorn. *Cf. supra* note 250.

[323] *qaran*,*ʿul*. It is native to India and Celebes (Ainslie **1**: 75–77). Its main use is as an aromatic. The Persian name is *khīrī*. The Arabic probably comes from the Greek. Ducros (184) gives *qaranful* today as *Caryophyllus aromaticus* L., a carminative, aromatic, and condiment. Clove is one of the minor aromatics (Māsawaih 403) good for a "death salve." Al-Kindī (90) used it so. Bayān (22) uses it for palpitation in heart preparations. I. B. (1748) declares that it is good for the stomach, heart, liver, and womb. *Cf.* also *Tuḥfat al-aḥbāb* (351).

[324] *qusṭ*. *Qusṭ* is the Arabic form of the Greek which may have come from Syr. *kusṭha* or *kustumbari* "coriander" (from Heb. *kusbar*). (*Cf.* Carnoy, 94). The latter is not likely. It is probably from *kushtam* (Sansk.). Diosc. (I: 16) lists costus, *Saussura lappa* (Dec.) C. B. Clarke perhaps, as good for convulsion, in ointments for the ague, for paralysis, and as an antidote. Māsawaih (407) used it as an incense. Al-Kindī (94) uses costus in *khalūq* salves. Maim. (338) gives its synonym as *al-bustaj*. In I. B.

kneaded, dried, and hidden. The way to know them is to throw one into water. It is left for an hour. If it is falsified, it is dissolved; if it remains in its condition, it is good.

Ginger.[325] If it is desired to distinguish in the domesticated ginger, the green and the dry types, then a little of it is chewed. If its warmth tends to bitterness, then it is the dry. If its warmth is sweet, then it is the Chebulic type. If it is desired to know whether the domesticated Chebulic type is the green or the dry, then the seed is broken up. If there is black inside the seed, it is the green type; if otherwise, then it is the dry. If its seed dissolves, it is the green. The taste of the green is good without bitterness. Somebody has said that the seed goes to pieces with a needle. This is not correct.

The water of the water lily.[326] The good part of it retains the pure white, good odor like the odor of the yellow water lily in summer. In its taste, it is obviously sweet and the fat is apparent. It tends toward being soft and weak. The falsified is the opposite of that.

Oil of *nārijil*[327] is oil of the coconut. The good and the falsified can be distinguished. If it is free from fraud, it becomes hard in winter time and its odor is good. That falsified with *halib* does not become hard; its odor is less than the good one. Each has its own shape. One is good and the opposite is bad. The latter should not be used.

Note on the testing of saffron of which there is the Genovese which is western. Of the western, there is a type known as the "oiled." It is said that it is found in its place in this state. **(90)** And it is also said that it can be sprinkled with oil to make it oily. The first statement is more preferred. One kind of it is "honeyed." It is said that it is sprinkled with honey. As far as I am concerned, this is impossible. Its value is less than the good, the pure Genovese, that is the western. Its crack is red to white as if it had been made rotten. Its crack on its upper end is thick; the lower is pointed. If it is dried, it rubs quickly. When it is chewed, it is bitter in taste and somewhat astringent. It burns the tongue a little. Its odor is good and it dyes strongly. Its yellow is toward red and it is light of weight. When it is dried in the sun, it dries quickly. The falsified is heavy in weight and its crack is even. When it is dried in the sun it contracts and becomes soft. It can be kneaded under the hand as long as it is warm. When it is removed from the sun and it is left in the air, it dries; then there is not in it the brilliance of the pure kind. When it is ground, a bit of it remains in the millstone; some of it is stuck. Its yellow tends toward white. Its odor is weak. As to the North African, it is imported in the form of discs. The pure part of it is light in weight in relation to the bad part. Its color is red to yellow. It is distinguished quickly. Its odor is good and strong. It is ground to a red color; the bad part is the opposite of this. The good Iraqi is light in weight; its crack is thin and the end is round. It tends toward white as if there were in it something spoiled. Its odor is stronger than that of the Genovese. Then there is the Karaki. It is inferior and there is little of it. It is not tested and it is not used. The way to bring forth its falsification is by melting it and leaving it in a piece of cloth. **(91)** Then there remains a little of it. It does not leave a coarseness in the mill. If some of it is used to dye, it dyes tending toward green. It is not a popular dye. Its odor is weak. It is dissolved by water and left. If a precipitate settles from it, then it is falsified.

Sapanwood is of two kinds, the Socotran and the other one. The Socotran type is the hard one and can be falsified. It is known that the Socotran is light, bright where broken, tasteless, and strongly red when pulverized. The falsified is the opposite of that.

The "eyelense of the cow"[328] is light. It is peeled one

(1785), Leclerc says *qusṭ* may resemble the present day *Aucklandia costus*. In Arabic times, it was used in impetigo, quartan fever, alopecia, and in other ailments. The *Tuḥfat al-aḥbāb* (350) states that the most estimable type is the white. In literary language, this word is written with a *kāf*.

[325] *zanjabīl*. In ancient Mesopotamia, ginger was probably in Sum. *KUR.GI.RIN.RA.ŠAR*, and in Akk. *kurkanū* (*DAB*, 158). It was used for the eyes, muscles, fumigation, and as a stomachic. Diosc. (II: 160) mentions it, *ζιγγίβερι*, as a condiment, good for the stomach and belly, for the eyes, and as an ingredient of antidotes. In Arabic times, *Aqrābādhīn* (115a) has it used for worms. Ginger (I. B., 1125) was used in the bath, as collyrium, as an aphrodisiac, and stomachic. It was used, essentially, in the same way in India (Ainslie 1: 152–153). Bayān (22) employs ginger in a medicine to improve the memory. *Zanjabīl shamī* is used today. It is grown in Iran and Iraq. However, this species is the resin described in Maim. (353). It is the *Inula Helenium* L. *Zanjabīl*, according to Ducros (117) is the *Zingibir officinale* Rosc., still used as a condiment, aromatic, aphrodisiac, and excitant. In North Africa it is *skenjbīr* (*Tuḥfat al-aḥbāb*, 143). The Greek word comes from the Pali which in turn comes from a corruption of the Sanskr. *çṛñgavera* from *çṛna* "horny matter" because the bracteates of ginger resemble "horny matter" (Carnoy, 275).

[326] *nailūfar*. The water lily or nenuphar is probably originally from India. *Nailūfar* or *nilufār* comes from the Persian *nīlūpar* which is from the Sanskrit *nīlōtpala* "blue lotus." *Nailūfar* indicates both the blue and white varieties, *Nymphaea caerulea* L. and *N. lotus* L. var. *alba*. Diosc. (III: 132) mentions the yellow and white types. The white water lily, *νυμφαία*, is effective for dysentery, for the spleen, stomach, bladder, and with pitch for alopecia. The lotus, is described in Diosc. (IV: 113). The Arabs (*cf.* I. B., 2243) thought the water lily good for insomnia due to heat, for pleural affections, and the lungs. The *Tuḥfat al-aḥbāb* (288) states that *nīlūfar* is also called *al-tājir* "the trader" since it opens its flower by day and closes it by night. The water lily is found today in Mesopotamia, Europe, Siberia, and India (Hooper, 144). Ducros (38) states that the flowers of white Nymphaea, originally from temperate Asia, are used as a refresher.

[327] *nārijil*. The coconut from *Cocos nucifera* L. was unknown to the Babylonians, Egyptians, and Greeks. It is found in the East Indies. The Arabic word comes from the Perisan *nārjīl* which is from Sansk. *nārīkeli*. Maim. (257) states that it is the nut of the Indies, *jawz al-hind*, and that it is also called *al-nāranj* or *al-rānaj*. The latter, of course, are mutilations of *nārijil*. The *Tuḥfat al-aḥbāb* (286) editors, Renaud and Colin, believe that the first mention of the coconut is to be found in Serapion, end of the ninth century.

[328] Uncertain.

layer after another. Its taste is bitter. In the inside of the "eye lense" is a black hard thing. The nearest to it is probably coagulated blood. The skin is smooth, soft in the rubbing. The falsified is the opposite of that.

Laudanum.[329] If it is desired to distinguish the falsified from the true, the latter is soft, tasteless tending toward astringency; its greenness is not bad. It is light in weight. When it is chewed, no sharpness is found under the teeth; there is no residue. The falsified is the opposite of that. Its odor is one already obtained; its recollection is not strong.

Note on testing of aloes.[330] The types are of Socatra, Medina, Waᶜza, and Hadramut. The Socatran is the best; its test is that, if it is quickly rubbed and has no odor, it is good. If one breathes over it, the appearance of its color is as the color of the liver inside. When it is rubbed, its rubbing is yellow to saffron. It is in between lightness and heaviness. The Medina type is close to it except that its rubbing is green tending to a yellow. The Waᶜza is bad in odor, easy to break, medium in a mild heaviness, easy to rub except that it turns to green. The Hadramut type is the most inferior of all. There is in it black tending toward green; it is bad in odor, heavy in weight. It is not used except to dye wool or ink of other things.

Note on the testing of blue bdellium.[331] It can be falsified with what remains of myrrh. It comes from that which was not extracted in a basket from its tree. Its test is that if it is buried its odor is smelled; it is something like the odor of myrrh. The broken part of it is hard, red in color, bereft of odor. Its color is toward blueness where broken, medium in heaviness. The broken part is smooth. Its taste is a little astringent.

Note on the testing of myrrh oil. It is called storax,[332] a liquid. It is said that it is like the fluid of the crushed mulberry. The color of the good part of it is white to blue, its odor strong without taste. If there is some oil, it remains gelid. It does not become dry. If it is taken between the fingers, it does not stick. The falsified is the opposite of this. Know it. The yellow amber of it is red tending to white. The test of the genuine is that, if it is warmed with rubbing on a cloth, and then brought near straw, it is attracted. It can be falsified with sandarac. In that case, the sandarac when broken is smooth and it is blue. If a bit of it is put in the fire, then its odor is smelled. It is similar to the odor of the "mastic[333] gum of the lentisk tree." Its yellow is deep in it while ambergris does not have a good odor.

Note on the testing of henbane.[334] If it is desired to know the genuineness or falseness, it can be seen from each of the solutions. It can be dyed with white lead and it is sold. Its whiteness comes out when one washes it more than once with water; the white part is dissolved. It is dried in the air and its color, red, returns. It has an extremely good odor. In it is put charcoal of wine. The quince and the charcoal are allotted equally. It is improved when you take it and make a wick of it. If it is desired to perfume with it, the fire is lit, then it is extinguished to make a good smoke. I have seen a small piece or a bigger one made out of myrobalan, and cut. It is left in the container; the smoke emerges from a man's house. It is good and pleasing.

---

[329] *lādan* or *lādhan*. The Babylonians knew laudanum as *ladunu* (Meissner I: 243). It is the resin of the Aegean plant, *Cistus ladaniferus* L. or another species of *Cistus*. The Greek (*cf.* Diosc. I: 97) name, *λάδανον*, comes from the Semitic (Carnoy, 156–157). Mishnaic Heb. = *lotem*. It is one of the secondary aromatics in Māsawaih (408). Laudanum was used by al-Kindī in a *ghāliya* recipe. Maim. (208) states that it is the resin, *dibq*, extracted from the plant called in Greek, *κίσθος*. I. B. (1999) mentions *ladan* to calm pain with rose oil on the fontanels of an infant, for the stomach, and for tumors. The *Tuḥfat al-aḥbāb* (241) states that *ᶜanbar* is equivalent to it. This is not correct. *Cf.* note 115.

[330] *ṣabr*. Used extensively in Arabic medicine. Bayān (30) considered it effective for biliousness.

[331] *muql azraq*. A resin from the *kūr* tree.

[332] *maiᶜa*. Storax. Maim. (228) mentions two types, liquid storax = *maiᶜa sāi'la* = *al-lubnā* and solid storax = *maiᶜa jāmida* = *iṣṭurk* = *ṣaṭrākhī*. Among the Arabs, there was much confusion as to what storax was. This is discussed in I. B. (2196) in a note by Leclerc. *Cf.* *Tuḥfat al-aḥbāb*, (58).

[333] *mastaka*. Probably the mastic resin from the *Pistacia Lentiscus* L. (note by Leclerc in I. B., 2139). It was employed in the Greek materia medica for the stomach. Its odor was also held desirable (Diosc. I: 17 *σχοῖνος*). However, it was probably known in Babylonian botany. The latter study is still undeveloped. In Arabic times (I. B., 2139), it was also used as a stomachic, for obstructions, and to combat nausea. The Greek word, *σχοῖνος*, from Diosc. is of Indo-European origin (Carnoy, 236).

[334] *banj*. *Hyoscyamus niger, albus, aureus* L. (al-Ghāfiqī, 162) *Banj* is also known as *saikurān* (*Tuḥfat al-aḥbāb*, 77). It is to be found in the deserts of North Africa, Egypt, Persia, and India. *Saikurān* is related to *shakiru* in Akk. (^{šam}*GUR*$_{3}$ in Sum.) which is probably the *H. niger* L. It was used in Babylonia for swellings, the lungs, for a poultice on the neck, and for headache. *Shakiru* = Aram. *shakrōnā* (*DAB*, 230). Henbane seed (Ainslie 1: 167–169) in India was used to induce sleep and to keep the bowels open in the case of melancholy and mania. The Persian is *bunch*. Diosc. (IV: 68) mentions *hyoskýamas*, *ὑοσκύαμος*, for sleep, pain, inflammations, and gout. I. B. (356) states that the Arabs used henbane for essentially the same purposes.

APPENDIX

Translation of Abū'l-ᶜAbbās Aḥmed ibn Muḥammed al Sufyānī

Ṣināᶜat tasfīr al-kutub wa-ḥill al-dhahab

(Art of Bookbinding and of Gilding) from the Arabic text published by P. Ricard, 1925.

1. *The introduction is omitted because of its irrelevant religious, and abstruse nature.*

2. *Chapter on the manner of making the cover boards.*

The cover board is the paper board which is covered by the leather of the book. To make it, you take a leaf of paper and smear it with starch paste. You leave it on your right. You smear a second leaf which is in front of it. Put the smeared side of the leaf on the smeared side of the second leaf and press it with your palms. Turn over the lower side to the upper. Notice if there are any wrinkles in it. If there are, then it is stretched. Make it flat with your palms until it is stretched to the utmost; not a wrinkle remains in the two surfaces and it is not soft. Then put down the two leaves which have been pasted to one another. Take two other leaves and do the same with them as you did with the previous two leaves until all the leaves are stuck face to face, two at a time. Spread them out in a warm place on the ground where there is no dust which might adhere to the pasted leaves. Wait until the boards dry. Then divide them according to the number of boards which you made. See how many are required of the number of leaves to be arranged to make the cover board. If you wish to make it thin, then subtract what you wish from the number of leaves. If you wish to make it thick, then add what you wish according to a measure that seems proper to you. After that, take what you gathered of the leaves for every cover board; for example, it may be five, six, or seven leaves according to your desire. Lay them around you. Take the first leaf and flatten it on a wooden board or stone surface. Smear it with paste. Put it on your right. Daub the second one and put it beside it. Daub the third, the fourth, and so on to the last. Every time one is daubed, it is left beside the one before it. After that, take the first and spread it out on the mentioned slab upon which you smeared the leaves. After you have flattened it on the table, take the leaf beside it which had been daubed just after it, and put one leaf on the other, i.e. the side daubed with paste to the side daubed of the other. You rub it with the pressing of your palms. After that, smear the upper dry side with paste also. Take the third leaf which had been previously daubed and put them together, smeared face to smeared face. Rub them with pressing. Smear also the dry side; put the fourth leaf on it in the manner that you have pasted one to the last leaf. When you have rubbed it with pressing, then take a leaf of the dry leaves and put it on the last leaf of the dry side. Rub on the dry leaf; rub it hard with a thick block, for example, a "form for smoothing." It is blunted at the edges. You rub it with its edge until the excess paste is extruded from between the pasted leaves. Then remove it. Put it down in a convenient place as the board or paper or what resembles that. Make another cover board; put one on another until you complete what you wish of the work of the cover board. Then take the prepared cover boards and put them between two thick boards of good wood. This is to exert pressure with the wooden boards after you have put sheets of paper between the cover boards. The size of the paper is larger than the two cover boards on the right, left, top, or bottom. The cover boards are tied strongly within the wooden press boards until you see water come out; this is from the paste with which the sheets were stuck together. Leave them in the wooden boards about a half or a whole day. Then, remove them from between the boards. Remove the paper which had been inserted between them. You will find them as you like and wish. Ask that grace may befall those who taught you. Then spread them in a place having warm air, not in the sun since the sun spoils the work. Leave them a night until morning breaks. In the morning, stand them up on their edges along the wall. When they are dry, then the utmost of goodness is attained especially if the paper is faultlessly good. There is no dampness in it due to water or moisture, or any deterioration in it. If there is no deterioration in it, it can take the rubbing after it has been covered with leather, until the side of it appears like a glass mirror.

3. *Chapter on how to tie the quires of the book, the pressing, the covering with leather, designing its center, how to work the headband*

The writer said that for whoever begins bookbinding after making the cover boards, there is a relationship, one to another, of the pages of the book. It appears on the bottom of the leaf and at the beginning of that which follows. This determines how the quires are completed or tied at their edges, when finished, by the order of their gathering. You are certain of the correctness and completeness of the book when all the quires are begun one with another. It is wrapped in covering leather like that found in the covered harem. After the leather has been put on the book, it is placed on a smooth stone to prepare it for the pounding. The leather is pounded with a heavy iron weighing six *raṭls*, or five or four. The purpose of that pounding is to make it even. The pounding is correct when the strokes are beside one another so that the book settles. Its paper is flexible so that some joins with the other according to the strength of the pounding. This is because the pounding does to it what the pressing boards do not. After that pounding, the wooden boards hold it firmly. If you work it in the wooden boards without pounding, then the wooden boards do not help a bit. The paper, then, does not ever settle together no matter what the press is. If you work the book in the boards after pounding, in whatever press you gather it, the paper obeys you. It is workable and flexible even if it is in the weakest of boards. Then you know this beneficial characteristic, O Bookbinder. After that, denote the middle leaf of the quires by a special instrument. Then all the quires are gathered leaning on their heads so that they are of the same related form and length, and all are in perfect register. If the writing on the sheets is higher or lower on some, then those sheets are moved up one on the other, the upper to the lower, into proper relation of the work to the desired condition in the circumstances. After that, two lines are drawn on the spine of the quires in the places where you will tie the book. You introduce the needle with the thread in the spine of the quire in the place marked with ink. The thread with which you bind is fine and strong, well formed, twisted on either three or four strands. The benefit of this is the achievement of the correct proportion; it is obvious to the bookbinder. The purpose is to gather the quires of the book by sewing them to each other. If there are many quires so that it appears thick where it is sewn, then pound it where the thread is with a mallet on a slab until the thickened thread is thinned out. It completes this part of the work.

Then the wooden boards hold the book in its thickness. A measure of the width of two fingers from the spine of the book protrudes. The quires are equalized as a group on their backs. Take care that the quires are lined up well. Near you there should be an iron tool like the curved blade knife of the shoemakers. With it you count the backs of the quires at one edge; then you record the number with certainty. Then you make the count on the other edge. If you find the number is the same, then know that the quires have not slid at all. If you find one number to be less or more, then examine the lesser side. Search for that amount by which the side was less until you find it. When you find it draw back the board a little bit. Introduce the awl in the middle of the slipped quire and move it gently until it matches its neighbors. When all of them are equal, then tie the board. Put the glue on the backs of the quires and spread it with your forefinger until it settles between the quires. Take the rubbing instrument in your hand. Introduce the tip of its end between the quires and let the glue flow between them with care but not deeply. Continue like that between the quires and those following until you come to the end.

After that, pass the forefingers over it until you are sure that the glue has been extruded from between them. I mean all the quires. Then loosen the press board and pull all the book into the center of the board until the quires are even with the edges of the board. Tie the board straight along the two sides. If the glue between the quires is excessive, it comes out and none remains except what is necessary for the operation. Wipe the backs of the quires with a tool until the excessive glue is removed. If something lumpy appears, then pound it lightly until it settles and equalizes. After that, let the board stand by the wall on your right. Then pare the two hinges of the flexible leather so that there is no stiffness in it. Fold the hinges on the limiting edge of the side of the book as you wish it, and according to the covering on the backs of the quires, i.e. the spine of the back.

If the book has leaves with gilt edges or edges which have colors made fast with gum arabic, then you may fear that you will damage the book while working on the two hinges. This may be caused by dampness from a trace of water so that the colors will run and have an odor and stick to each other. In that event, put the folded parts of the two hinges away from the sides of the book in such a manner so that the hinges do not come in contact with the writing. Thus it will not be harmed by the dampness. When you prepare the two hinges, both being wide, glue them to the book when they are dry, neither moist nor wet. If you wish to attach them both, then untie the press board. Loosen it carefully from the book, then pull the two hinges together with the two edges of the book. This is after the book has been set back a little from the front edge of the board. Tie the board firmly and evenly. Use an awl on the edges of the book until the glue is worked between the hinges and the book in a straight and even fashion. After that, do so from the spine of the book and the two hinges. Turn over the two hinges on it, each of them on the other with awling and flattening. Then smear three sheets with glue and put them on the spine of the book. Use the awl to help place them properly in regard to the hinges, the middle and the edges of the book. Let the board stand in temperate air a day and a night. When it becomes dry and hard, pass a sharp blade over the edges of the book and cut off the excess paper gathered with the glue still sticking to the spine of the book. Untie the board and introduce the awl between the book and the board to separate it from the board.

When you have removed the book, measure two cover boards on it. This is after you have gone around the edges with the cutters; it is also after you have come down on the edge of the lined cover boards with the cutting blade and cut them straight where they are to be fastened to the two hinges. Then put three drops of glue on the hinges, or four or five drops, according to the size of the book. Put the cover board on it. Do the same with the other side. Put the book with its two cover boards between two thick cheeks, bound with pressure on the boards. Leave the book between the two of them until the glue holding the cover boards to the two hinges dries. When it is dry, untie the book from the cheeks of the press; you will find it straight. Then using the divider, measure it (the book) correctly on three sides. Cut it from all sides; rub the cut part with fuller's pumice until the cutting trace of the iron has disappeared. Wipe with your palm whatever you have artfully done with the stone. Rub the stone on it; then you burnish it to the limit.

When you complete the cutting of it, take the measure of one-half the right cover board of the book having three cover boards. Cut a tongue cover from it for the written part of the book. It is called the tongue (*lisān*). From what remains of the other half of the cover board from which the *lisān* was taken, cut the fore-band. This is attached between the small cover board that is at the end of the book and the large band that is attached to the beginning cover board.

After that, carry on. Divide the front cover with a marker into two halves. Put the marker in the middle of the cover—it is the art of the oriental bookbinder—and make a line on it. After that, use the impressing tool according to the mark. Press it down evenly and be sure that the goffering tool comes down exactly on the marked line.

When you cover the front cover board with leather, rub it to the right and to the left. Take the cover boards off from the book and flatten them on a marble slab with your hands. Let the stamp come down from directly above the mark on the leather and tap it lightly with a small mallet. Do it gently so that the leather will not be cut. Repeat the tapping with the mallet on the tool. When the glue appears excessive in spots owing to the tooling, press the leather to the right and to the left until the excess is smoothed out. The end of the goffering tool is applied to the leather with force until, when you remove it from its place, the trace of its tip remains as a sharp edge just removed from wax. Repeated tapping on the goffer causes an embossment and it makes the imprint very smooth. When you complete the stamping, turn the leather over on the edges of the cover board.

When you complete the first cover board, flatten it on a marble stone with your hands. Put the book down on it in the condition it was in before the covering was applied. Attach the cover board temporarily. The leather may contract after drying and after the headband is sewn so that the thread of the headband adds bulk to the spine of the book. If you wish, fix the unbound book in the cover after you mark it, dry it, and line it inside. Then if you find that the binding is small on the book because of the sewn headband, let out the amount of that excess tightening from the binding.

If the book pages are short, and the bookbinder is thoughtful and intelligent, he knows what is too much or too little, what is fitting and not fitting. When you put the book down on the first cover board, daub the glue on the third cover board. Cover it with leather. Complete the work on it as on the first. Put the small band down beside it after daubing it with glue. A measure of one or two fingers or less is between it and the cover board. After that put the large band down on it after you have glued and rubbed it. Then apply the designing tool which is one-fourth the size of the large designing tool used in the middle of the first and second cover boards. Then, attach thinned leather to the edge of the other cover board and on the outer edge of the tongue. The small band is in the middle under the pared leather. You rub the work and decorate it.

Don't be disappointed in the leather; if there is looseness or bulging, then lessen it by rubbing when you cover the second cover board. You rub it a bit in the direction of the tool. Wrinkling in the decorated part disappears after pressing and rubbing with care.

4. *Chapter on covering with leather*

Hang it [the leather] on a reed or string in warm air but not in the sun because when the sun is on it, as we said before this, the work is spoiled. Leave it on the reed until morning. If, on inspection, you find it can stand rubbing, that is, if there is wetness and moisture in it, then it needs rubbing. Leave it until it dries even if it is a day or two or more. Then, when it is rubbed, it comes out softened as you wish it. If you desire, polish it more than that. Daub it with water. If it absorbs the water, then find someone to hold it on the slab while you rub it so that it does not slip on the stone. Rub it back and forth with an oyster shell or with a suitable wood burnisher made by the shaper. Then the results are good. If some of the decoration is spoiled by the vigor of the rubbing, then restore it by tapping the decorating tool with a mallet. It is thus returned to its old form as you want it. After you complete this aspect of bookbinding, you line it either with leather or cloth. Leave it to dry and work on the sewing of the headband.

How to attach the headband: There should be near you gum arabic dissolved in water like thick honey. From it, put some on the head of the quires on the edge under the strip on which the headband was sewn—in the manner that you put the thong on it. The strip itself is of tanned leather which had been smeared with gum arabic before that until dry. Start taking the leather strip from it when you need it for the headband. When that gum which you have put on the spine of the book is dry, then the leather strip is wetted with your saliva and put down on that place whereon is the gum arabic so that the two stick together by the glueing.

The needle with thread is inserted into the middle of the right quires after the end of the thread is made fast in the spine of the book in the back side from which place the head of the needle comes out. If you enter it in the middle of the other quire, go ahead in the same fashion with the sewing of the quire until you finish on the last quire. Fasten the thread well on the last stitch; the sewing is then inseparable. Complete the remainder weaving it with colored silk until you complete the work of the headband from the two sides. After that, fix the cover boards on the book after you have smeared it with glue. Tie on the spine side with strong thread. Put the book between two heavy tablets as was done previously. Press on them both with the board. Leave it between the two tablets, holding together and drying. Then you will find that it comes out straight as desired. Then exalted Allah leads you to success in the right way. May He be praised.

5. *Chapter on description of solution of gold, its washing, its soaking with glue, and description of writing with it*

This is after it is dried. Take a leaf of the gold with which they write, and rub it well until it is ready. If there is little gold or more, like one or two *mithqals*, rub it in a glass dish having a flat, deep bottom. When honey becomes doughlike, rub it with wood of a light type until it is well prepared. Pour water on it; stir it. Leave it for a short time. Pour water on it in another glass vessel; do so gently. Add other water to the gold; let it drip on the preceding first water. Repeat the pouring of the water and the dripping until the good of the honey comes out so that no sweetness is left in it. Then raise the vessel containing the gold onto hot ashes until it is dry with no moistness remaining in it. Then remove it and protect it from dust and insects which eat all they find with the smell of honey.

Let us return to discussion of the water you clarified from the gold. Leave it in the vessel overnight. When morning comes, you find that the gold which flowed with the water, is stuck to the bottom of the vessel on the glass. The water of the honey floats on top. Decant the water from the gold which is stuck. It is not moved. When you decant the water from it, grasp it between your fingers and add other water to it. After an hour, clarify it and let it drip into the vessel from which you write. It is a small glass vessel, pretty in appearance. Add to it what you wish of that dry dissolved gold, more or less, in the amount necessary and desired for writing. Work gum arabic into it in a measure that satisfies you, or fish glue if it is available. Soak it in water; make a wool *līq* and stir it with the pen. Write on paper what you wish with the pen. When it is dry on the paper, rub it with an oyster shell. Do not forget to shake the *līq* and turn it upside down. If you wish to write with it on leather, then do not put gum arabic into it. Put into it fish glue, especially. If you write with it, leave it until it dries. Rub it with the oyster shell or with something like it as you like and desire. Allah is the helper. When you are through with the gold soaked with the glue, then watch it. Do not leave the glue in it or else it becomes thick, worms are created in it and flies eat it on account of the odor of the thick glue. Clarify it once or twice until the glue odor does not remain. Remove it and keep it. We will give you a helpful note.

Note [*a*]. Know that the fish glue mentioned is of two types. One is yellow tending toward red. You dissolve it with water on a mild fire. The gold is soaked with it. Its preparation is from Achilles' tendon. It is cooked as they cook the strong glue of the leather. You have watched how this was made. The second glue is an uncooked glue remaining as it was. It can be described as old dried soap. It comes wrapped one on top of the other. Its color is ivory white. The method of preparation is that you take a measure of what is good for the gold work. You put it into water until it is moist. Then pound it on a marble stone. Fold it as one folds an amulet. Return, and beat it until it is stretched a second time; it becomes like a sheet of paper. Fold it also and beat it until it is stretched. Cut off a small piece and put it in a little water, as much as is enough for dissolution. Put it on a light fire. When it boils, it is dissolved. Take it down from the fire and rub it with your forefinger until it attains a dissolved gluelike viscosity. Add more water to it and return it to the fire until it becomes like boiled olive oil. Leave it until it grows cold. Soak the gold with it. Stir it and stir the *līq* in it. Try writing on the leather with it. When dry, rub it. When you see that its color is shiny, wipe it with your finger.

If the gold is wiped off, then know that it is because of too little glue. Then add more glue to it, as much as is proper. If you see that the gold adheres to the leather and its color shines, then that is the desired. If you rub it and see a disappointing color without any shine, then know that there is too much glue and that the leather has not absorbed the glue. That prevents it from shining. Therefore, add more water without glue to it. Warm it so that it melts and dissolves its glue. Then add a bit of water to it and clarify it. Then the glue of it is diminished until there remains in it as much as is useful to you. When you write with it, the leather absorbs the glue. Rubbing is useful to make the shine appear on it. It will not wipe off from the leather. When you wipe it, then this glue is better than the other glue. Only those men know it who have examined it and know its qualities. I shall give you a useful note.

Another Note [*b*]. In regard to the mentioned fish glue, everytime I recalled to you its cooking and the soaking of the gold in it, it meant in the warm season. As to the extremely cold period, if you soak the gold in it, it solidifies and does not flow in the leather. If you soak it with water and then put it on the fire, then it will flow. When it gets cold, it solidifies again until the gold color comes out from it. It appears to you that the water is important here. It solidifies and does not flow until it is placed above the air of that fire whose air is like the heat of summer, that is, the heat in the shade, not the heat in the sun itself. It is necessary for elegance that you hang the vessel of the gold containing the glue belonging to it over the fire in a fireplace. When the glue feels the heat underneath, it starts writing. Understand. Be intelligent and sharpminded. The period of the cold will tell you all that I have described to you while experimenting. The soundness of the thing I have related to you will be shown correct.

As to the shopkeepers, they do not know of the glue of the fish except the Syrian type which they have. As to this latter glue, I found it in possession of a man who knows it. He said that while the shopkeepers had it, the broker was prepared to sell it by the ounce. We described it to the man and another man in the same business who knew it. They agreed that they would buy it at that price and divide it between them. When I found it with that mentioned man he said, "I will not sell it except at a *raṭl* for an ounce." And I couldn't but take it at the price he asked, since I needed it. I started to take care of it and speak of it and be proud among those in the same business while they didn't know with what it was that I was ahead of them. There is a saying, "The utensil helps." It is said also, "The utensil is half of the work or half of the skilled worker." The exalted Allah is the director of truth. All of this treatment is with the glue of the white fish. However, for those who are contented with Syrian glue, then it suffices.

Note [*c*]. If you wish to cover with a certain leather you want to use, you have to be careful that you never use the leather originating from on top of the spine unless you wash it with water. This is because the tanners, when they dye the leather using alum, will get a dye color to shine more by permeating the leather with oil. Then, its color comes out to the limit. Brother, you need as advice, which is best, that to make the leather which you have cut as big as the cover of the book, let it pass through your hands while you rub it in the water. The oil rises to the surface of the water. This water is decanted. The washing and rubbing are repeated so that it frees the oil on the surface of the water. Repeat it until it weakens the oil in it. This is because if you cover the book before washing it, and you let the gold flow on it, then the oil hinders the leather from the absorption of the glue. Here, by the glue, I mean the glue of the fish. As to the Syrian glue, if you melt it, it has a special characteristic in a vessel. You leave it until it solidifies in it.

When you wish to use it in gold, put a little water on it, as much as you used to soak the gold. Rub it an hour with your index finger until it becomes like a white *līq*. Soak it with your gold. Write with it on leather, washed or not. The work will come out the best. Rubbing of the gold is also done without washing the leather. As to my warning to you about washing the leather when you soak it with fish glue, it is because, if you rub it, the gold will peel off from the leather but this glue is good even if you do not wash it.

If you wish to dye leather a raisin wine color and then the dye comes out spotted, it is because the oil keeps it from some places. If you wash the oil from it and cover the book with it, the color of the leather will disappear because of this. However, if you rub it, and repeat the rubbing, then you will get the color which the tanner did not because of the existence of oil in it. If you wish, dye the leather a raisin wine color after you wash it with water. Press it well and stretch it well to prevent it from wrinkling. Dissolve a bit of vitriol in water and smear the leather with it. Not too much is needed when you have applied water to the leather. When it appears that it needs a darker color than that, repeat the smearing until its color is satisfactory to you.

However, if there is little water, you may be afraid that it will disappoint you and darken the color on the first smearing so that it differs from that which you desire. But, if the water is abundant, then smear it time after time until its color satisfies you and it does not disappoint you. When you complete the dyeing on the leather, then introduce it into water and wash it well so that the dye does not streak it. It darkens its color. If you wash it, it prevents any further coloration. The author says, "This is the last of what has occurred to my memory while writing. This is in the month *dhu al-hijjah* in the year A.H. 1029.

Description of dyeing leather with violet. You take leather which has been tanned with white beam-tree. Wash it well with water. Pare it with an iron tool until no dirt remains on it. Sew it together until it is swollen like a leather sac. After that, soak it with an ounce of alum dissolved in water. Blow into it and rub it until the alum water goes into it. Empty the water on it. After that, soak it with water in which the good species of sappanwood (*baqam kaḥāl*) has been cooked. This has a good taste on the tongue. Then blow into it after soaking. When it swells, turn it in your hands. Turn it upside down, its bottom to the top and its top to the bottom. Open the mouth of the leather sac; it will be satisfactory to you. If not, continue the soaking until its color will please you.

6. *Chapter on the description of decorating the leather for binding*

You take the pared skin and smear it with strong glue on both sides. Then you place two unpared skins on it, smeared with glue on the inner side. Leave it until it dries. Apply on it a sheet of paper on which is marked with ink any design which you wish to use. This is done by taking a sheet of thin paper, wetting it with your saliva, leaving it until the saliva is absorbed, and dries a little. Press that sketch, whatever it is you wish to draw—a design, an illustration, or anything else relating to the book—with your thumb and your finger. Outline the sketch on it, remove it, and leave it until it dries. Outline it with pen and ink until the design is visible. When you have glued it on the previously mentioned leather, and it has dried, follow the sketch and the impression of the described drawing with a *mubazif* like that of the blood leather. To decorate make the impression on moistened leather to reproduce the design. If there are any blank spots in the work, fill them in. Try other leather also until the work pleases you. I finished this on the twenty-third day of *Shiwal*, the more holy, in the year 1255 A.H. So be it.

The intelligent ones will understand this with simple directions. For others loud shouting will be necessary. Another group will need cursing but not the stick. A stick will be necessary for the last group.

SELECTED BIBLIOGRAPHY

The names or initials printed in bold-faced type before each item are abbreviations by which, for convenience, works cited are indicated in the text.

Achundow. Achundow, Abdul Chalig, *Die pharmakologischen Grundsaetze (Liber fundamentorum pharmacologiae) des Abu Mansur Muwaffaq bin Ali Harawi*, Historische Studien aus dem pharmakolog. Institut der Kaiserl. Universitaet Dorpat, **3**, Halle, 1893.

al-ᶜAlamī. ᶜAbd as-Salām b.M. al-ᶜAlamī, *Ḍiyā an-nibrās fī ḥall mufradāt al-Anṭākī bi lughat Fās*, Fès, 1318 H.

al-ᶜAlmāwī. ᶜAbdalbāsiṭ b. Mūsā b.M.b.Ism. al-ᶜAlmāwī ash-Shāfiᶜī, *Al-muᶜīd fī ādāb al-mufīd wal-mustafīd*, MS Damascus, 1349.

al-Anṭākī. Dāwūd al-Anṭākī, *Tadhkirat ūli al-albāb wa'l-jāmiᶜ, li'l-ᶜajab al-ᶜujāb*, Le Caire, 1281 H.

Alaune. Ruska, J., ed., transl., *Das Buch d.Alaune u.Salze*, Berlin, 1935.

al-Aṣmaᶜī. ᶜAbd al-Malik al-Aṣmaᶜī, *Kitāb an-nabāt wa'sh-shajar*, Beirut, 1908.

al-Damīrī. Kamāl ad-Dīn ad-Damīrī, *Kitāb ḥayāt al-ḥayawān al-kubra*, 2 v., Le Caire, 1309 H.

al-Ḥamdānī. Abū M. al-Ḥasan b.A.b. Ya'qūb al-Ḥamdānī b. al-Ḥā'ik, *Ṣifa jazīrat al-ᶜarab*, ed. D. H. Miller, Leiden, 1884–1891.

al-Ḥawī. Abu Bakr M.b.Zakariyya ar-Razi, *Kitābu'l ḥawī fī'ṭ-ṭibb*, 4 v., Hyderabad, 1955–1957.

al-Idrīsī. M. al-Idrīsī, *Maghrib, sūdān, miṣr, wa-al-andalus*, ed and transl. R. Dozy and M. J. de Goeje, Leyde, 1866.

al-Iṣfahānī. Ḥamza b. al-Ḥasan al-Iṣfahānī, *Kitāb ta'rīkh sinni mulūk al-arḍ wa'l-anbiyā'*, Berlin, 1340 H.

al-Jaubarī. ᶜAbd al-Raḥīm b. ᶜUmar al-Dimashqī al-Jaubarī, *Kitāb al-mukhtār fī kashf ul-asrār wa-hatk al-astār*, Cairo, 1908.

al-Kāmilī. Manṣūr b. Baᶜra al-Dhahabī al-Kāmilī, *Kashf al-asrār al-ᶜilmīya bidār adh-dharb al-miṣrīya*, MS Kairo V, 390.

al-Kindī. Ya'qūb b.Iṣhāq al-Kindī, *Kitāb kīmiyā' al-'iṭr wat-taṣᶜīdāt*, ed., transl. K. Garbers, Abh. f.d.Kunde d.Morgenlandes, **30**, Leipzig, 1948.

al-Majūsī. ᶜAlī ibn al-ᶜAbbās al-Majūsī, *Kāmil aṣ-ṣinā'a aṭ-ṭibbiyya*, Būlāq, 1294.

Alphita, R. Creutz, Das mittelalterliche medizinisch-botanische Vokabularium "Alphita," *Quellen und Studien z. Gesch. d. Naturwiss. u. d. Med.* **7**: 1–80, 1940.

al-Qalqashandī. Aḥmad ibn ᶜAbd Allah al-Qalqashandī, *Ṣubḥ al-aᶜshā fī ṣināᶜat al-inshā'* **2**, Cairo, 1913.

Al-qāmūs. *Al-qāmūs al-muḥīt*, Abū Ṭāhir Majd ad-Dīn M. b. Yaᶜqūb al-Fīrūzābādī ash-Shīrāzī, 2 v., Būlāq, 1272 H.

al-Qāshānī. Abū al-Qāsim ᶜAbd Allah b. ᶜAlī b. M. b. abī Ṭāhir al-Qāshānī, *Kitāb jawāhir alᶜarā'is wa-aṭāyib al-nafā'is.*, ed. J. Ruska *et al.*, Istanbul, 1935.

al-Rammāḥ. Ḥasan al-Rammāḥ Najm al-dīn al-Aḥdab, *Kitāb al-furūsīya wa-al-munāṣab al-ḥarbīya*, extracts in J. T. Reinaud and I. Favé, *Histoire de l'artillerie*, part I, Paris, 1845.

al-Rāzī 1. Ruska, J., *Al-Rāzī's Buch Geheimnis der Geheimnisse, Quellen u. Studien z. Gesch. d. Naturwiss. u.d.Med.* **6**, Berlin, 1937.

al-Rāzī 2. Abū Bakr Muḥammad ibn Zakariyyā' al-Rāzī, *Kitāb manāfiᶜ al-aghdhiya wa-dafᶜ madhārrha*, Le Caire, 1305 H.

al-Razzāq. Leclerc, L., *Kachef er-romoûz d'Abd er-Rezzaq ed-Djezairy . . .*, Paris, 1874.

al-Sufyānī. Abu'l-ᶜAbbās Aḥmad ibn M. al-Sufyānī, *Sināᶜat tasfīr al-kutub wa-ḥill al-dhahab*, ed. P. Ricard, Paris, 1919.

al-Tauḥīdī. A. b. M. b. A. at-Tauḥīdī aṣ-Ṣūfī a. Ḥaiyān, *R. fī ᶜilm al-kitāba*, MS Krafft 11, H. Krafft, Die arab. pers. und türk HSS d. k.k. Orientalischen Akad. zu Wien, Wien, 1842.

Anawati. Anawati, G. C., *Ta'rīkh al-ṣaidalah wa-al-ᶜaqāqīr*, Cairo, 1959.

André. André, J., *Notes de lexicographie botanique grecque*, Paris 1958.

Aqrābādhīn. Ya'qūb ibn Iṣḥāq al-Kindī, MS Aya Sofya 3603 Translated by M. Levey.

ᶜArib b. Saᶜd. *Ṣila ta'rīkh aṭ-Ṭabarī*, de Goeje, M. J., ed., Leiden, 1897.

Aristotle's lapidary. Ruska, J., transl., *Das Steinbuch des Aristoteles* Heidelberg, 1912.

Avicenna. Holmyard, E. J. and Mandeville, D. C., *Avicenna's De Congelatione Et Conglutinatione Lapidum, Sections of the Kitāb al-shifā'*, Paris, 1927.

ᶜAwwām, *Kitāb al-filāḥa*. transl. J.-J. Clement-Mullet, Paris, 186407; ed. J. Antonio Banqueri, Madrid, 1802.

Björkman. Björkman, W. and Kühnel, E., Kritische Bibliographie. Islamische Kunst 1914–1927 . . ., *Islam* **17**: 183–198, 1928.

Bosch. Bosch, G. K., *Ars Orientalia*, **4**, 1–13, 1961.

Brockelmann. Brockelmann, Carl, *Lexicon Syriacum*, *Halis* Saxonum, 1928.

Carnoy. Carnoy, A., *Dict. Étym. des Noms Grecs de Plantes*, Bibliothèque du Muséon, **46**, Louvain, 1959.

CAS. Thompson, R. C., *The Chemistry of the Ancient Assyrians*, London, 1925.

Colin. Colin, Gabriel, *Abderrezzāq al-Jezāiri un Medecin arabe du XIIᵉ S. Heg.*, Montpellier, 1905.

DAB. Thompson, R. C., *A dictionary of Assyrian botany*, London, 1949.

DACG. Thompson, R. C., *A dictionary of Assyrian chemistry and Geology*, Oxford, 1936.

Diehl. Diehl, Edith, *Bookbinding its background and technique*, New York, 1946.

Dillmann. Dillmann, A., *Lexicon linguae aethiopicae cum indice latino*, Lipsiae, 1865.

Diosc. Wellmann, Max, ed., *Pedanii Dioscuridis Anazarbei De Materia Medica Libri Quinque*, Berolini, 1907–1914.

Dozy. Dozy, R., *Supplément aux dict. arabes*, Leyde, 1881.

Ducros. Ducros, M. A. H., Essai sur le droguier populaire arabe de l'inspectorat des pharmacies du Caire, *Mém. prés. à l'Inst d'Égypte* **15**, Le Caire, 1930.

Fihrist. Ibn al-Nadīm, *Kitāb al-fihrist*, eds. Flügel, G. and J. Roediger, **1**, Leipzig, 1871.

Forbes. Forbes, R. J., *Bitumen and petroleum in antiquity*, Leiden, 1936.

Forskål. Forskål, Petrus, *Flora Aegyptiaco-Arabica*, Hauniae, 1775.

Freytag. Freytag, George W., *Lexicon arabico-latinum*, 4 v., Halis Saxonum, 1830–1837.

Frauberger. Frauberger, H., *Antike und frühmittelalterliche Fussbekleidungen aus Achmim-Panopolis*, Düsseldorf, n.d.

GAL. Brockelmann, Carl, *Gesch. d. arab. Litteratur*, Leiden, 1938–1943.

Galen. Kuehn, W., ed., *Claudii Galeni opera omnia*, Leipzig, 1821–1833.

Ghāfiqī. Meyerhof, M., and Sobhy, G. P., *The abridged version of "The Book of Simple Drugs" of Aḥmad ibn Muhammad al-Ghāfiqī by Gregorius Abu'l-Faraj (Barhebraeus)*, Cairo, 1932–1940.

Goldziher. Goldziher, I., *Vorlesungen über den Islam*, Heidelberg, 1925.

Gratzl. Gratzl, E., Creswell, K. A. C., and Ettinghausen, R., Bibliographie d. Islamischen Einbandkunst, 1871–1956, *Ars Orientalis*, **2**: 519–540, 1957.

Gunther. Gunther, R. T., ed., *The Greek herbal of Dioscorides*, New York, 1959.

Holmyard. Holmyard, E. J. *Kitāb al-kanz al-afkhar wa-al-sirr al-a'ẓam fī* taṣrīf al-ḥajar al-mukarram, Paris, 1923.

Hooper. Hooper, D., and Field, H., Useful plants and drugs of Iran and Iraq, *Field Museum of Natural History, Bot. Series*, Publ. 387, Chicago, 1937.

Huart. Huart, Cl., *Les Calligraphes et les Miniaturistes de l'Orient Musulman*, Paris, 1908.

Hyrtl. Hyrtl, Joseph, *Das Arabische und Hebräische in der Anatomie*, Wien, 1879.

I.B.. L. Leclerc, transl., *Traité Des Simples par Ibn El-Beithar*, Notices et Extraits des Manuscrits de la Bib. Nationale . . ., **23, 25, 26,** Paris, 1877, 1881, 1883.

Ibn Bādīs. Mu'izz b. Bādīs, *'Umdat al-kuttāb wa'uddat dhawi'l-albāb*. MS, Oriental Institute, University of Chicago.

Ibn Biklarish. Yūsuf b. Iṣḥāq ibn Biklarish, *Al-musta'īnī fi'ṭ-ṭibb*, MS no. 481, Bib. Rabat, Morocco.

Ibn Iyās. Abū'l-Barakāt M. b. A. b. Iyās Zain ad-Dīn an-Nāṣirī al-Čerkesī al-Ḥanbalī, *Kitāb ta'rīkh miṣr*, Būlāq, 1311 H.

Ibn Jamā'ah. Badr ad-Dīn a. Al. M. b. Burhān ad-Dīn a. Iṣḥāq Ibr. b. Sa'dallah b. Jamā'ah al-Kinānī al-Ḥamawī ash-Shāfi'ī, *Tadhkirat as-sāmi'ilkh*, ed., M. Hāskim an-Nadwī, Ḥaidarābād, 1353 H.

Ibn Sīnā. Abū 'Alī al-Ḥusain ibn Sīnā, *Kitāb al-Qānūn fi'ṭ-Ṭibb*, Būlāq, 1294 H.

'Isā. Aḥmed 'Isā Bek, *Tārīkh al-nabāt 'ind al-'arab*, Cairo, 1944.

Jābir. Siggel, A., *Das Buch der Gifte des Ğābir ibn Ḥayyān*, Akademie der Wissensch. u. d. Literatur Veröffentl. der Oriental. Kommission, **12,** Wiesbaden, 1958.

Kōhēn. Abū'l-Munā ibn Abī Naṣr al-Isrā'īlī, *Minhāj al-dukkān*, Būlāq, 1287 H.

Krauss. Krauss, Samuel, *Talmudische Archäologie* **3,** Leipzig, 1910–1912.

Kremer. von Kremer, A., *Culturgeschichte des Orients unter den Chalifen*, Vienna, 1877.

Krenkow. Krenkow, F., The use of writing for the preservation of ancient Arabic poetry, in *A volume of oriental studies presented to E. G. Browne*, Cambridge, 1922.

Lane. Lane, E. W., *Arabic-English lexicon*, London, 1863–1893.

Leclerc. Leclerc, Lucien, *Histoire de la médicine arabe*, Paris, 1876.

Le Strange. Le Strange, G., *Baghdad during the Abbasid Caliphate*, Oxford, 1901, 1923.

Levey. Levey, M., *Chemistry and chemical technology in ancient Mesopotamia*, Amsterdam, 1959.

Levy. Levy, J., *Neuhebraisches und Chaldaisches Wörterbuch über die Talmudim und Midrashim*, Leipzig, 1876–1889.

Lisān al-'arab. By Jamal ad-Dīn ibn Manẓūr, Būlāq, 1300–1304 H.

Loew. Loew, I., *Die Flora der Juden*, Veröffentlichungen der Alexander Kohut Memorial Found., vols. **2–4, 6,** Wien, 1924–1934.

Löw. Löw, L., *Graphische Requisiten u. Erzeugnisse b.d. Juden*, Leipzig, 1871.

Lucas. Lucas, A., *Ancient Egyptian Materials and Industries*, London, 1945.

MAA. Grapow, H., von Deines, H., Westendorf, W., *Grundriss d. Medizin d. Alten Ägypter* IV/**2,** Berlin, 1958.

Maim. Meyerhof, Max, ed., transl., *Sharḥ Aṣmā' Al-'Uqqār, Un Glossaire De Matière Médicale Composé Par Maïmonide*, Mém. Prés. *a l'Institut d'Égypte*, **41:** Le Caire, 1940.

Māsawaih. Levey, Martin, Ibn Māsawaih and his treatise on simple aromatic substances: Studies in the history of arabic pharmacology I, *Jour. Hist. Medicine* **16,** 1962.

Mieli. Mieli, A., *La Science Arabe*, Leiden, 1938.

Müller. Müller, Joel, *Masechet Soferim*, Leipzig, 1878.

Munshi. Qadi Aḥmad ibn Mir Munshi, *Calligraphers and Painters ca. A.H. 1015/A.D. 1606*, transl., T. Minorsky, Washington, 1959.

Nabat. Post, Geo. E., *Nabāt sūriah wa-falasṭīn wa-al-qiṭr al-miṣrī wa-bawādīkā*, Beirut, 1884.

Paradies. Siggel, A., Das Paradies der Weisheit u.d. Medizin, Abū Ḥasan 'Alī b. Sahal Rabban aṭ-Ṭabarī, *Quellen u. Studien z. Gesch. d. Naturwiss. u. d. Medizin* **8:** 216–265, 1941.

PR. Ebeling, E., Parfümrezepte und Kultische Texte aus Assur, *Orientalia*, **17:** 129–143, 299–313, 1948; **18:** 404–418, 1949; **19:** 265–278, 1950.

Ray. Ray, P. *History of chemistry in ancient and medieval India*, Calcutta, 1956.

Regemorter. van Regemorter, Berthe, La Reliure des Manuscrits Gnostiques Decouverts a Nag Hamadi, *Scriptorium*, XIV/2: 225–234, 1960.

Reinaud. Reinaud, M., *Relation des Voyages faits par les Arabes et les Persans dans l'Inde et à la Chine* (*Salsalat al-tawārīkh*), Paris, 1845.

S. Supplement to GAL.

Schweinfurth. Schweinfurth, Geo., *Arabische Pflanzennamen aus Aegypten, Algerien und Jemen*, Berlin, 1912.

Serapion. Guiges, P., Les noms arabes dans Sérapion "Liber de simplici Medicina . . ." *Jour. Asiatique*, May, 1905.

Sontheimer. von Sontheimer, J., transl., *Grosse Zusammenstellung über die Kräfte der bekannten einfachen Heil- und Nahrungsmittel von . . . Ebn Baithar*, Stuttgart, 1840–1842.

Steingass. Steingass, F., *A comprehensive Persian-English dictionary*, London, 1892.

Tāj. M. Murtaḍā az-Zubaidī al-Ḥanafī, *Tāj al-'arūs min jawāhir al-qāmūs*, Būlāq, 1307 H.

Tedhkira. Colin, G., *La Tedkirà d'Abū'l-'Alā', Publ. de la Faculté des Lettres d'Alger*, **45,** Paris, 1911.

Theoph. Hort, Arthur, *Theophrastus' Enquiry into Plants.*, v, 2 London, 1916.

Tuḥfat al-aḥbāb. Renaud, H. P. J. and Colin, G., *Tuḥfat al-aḥbāb, glossaire de la matière médicale marocaine, Publ. de l'Inst. des Hautes Études marocaines*, **24,** Paris, 1934.

Uṣaibi'a. Muwaffaqaddīn a. 'l-'Abbās A. b. al-Q. b. a. Uṣaibi'a b. Ḥalīfa as-Sa'dī al-Ḥazraji, *'Uyūn al-anbā' fī ṭabaqāt al-aṭibbā'*, Mueller, A., ed., Le Caire, 1882.

Vullers. Vullers, J. A., *Lexicon persico-latinum etymologicum*, Bonnae, 1855–1867.

Wehr. Wehr, H., *Arabisches Wörterbuch f. d. Schriftsprache d. Gegenwart*, Leipzig, 1952–1959.

Wiedemann. Wiedemann, E., Beiträge z. Geschichte der Wissenschaften, *Sitz. d. Physikalisch-medizinischen Soz. in Erlangen* **32,** 1914; **49,** 1918; **51,** 1918.

Yāqūt. Wuestenfeld, F., *Jacut's geographisches Woerterbuch*, 6 v., Leipzig, 1866–1870.

Zahrawi. Hamarneh, Sami K., *Some pharmaceutical aspects of Al-Zahrawi's Al-Tasreef about 1000 A.D.* Univ. of Wisc. thesis, 1959.

Zakī. Zakī M. Ḥasan, Al-kitāb fil funūn al-islamīya, *Al-Kitāb* **1:** 252–263, 1946.

to integrate (gold in honey to make liquid gold)	ابتلع
needle	أبر
pure gold	ابريز
ebony	ابنوس
citron	اترج
furnace	أتون
glass furnace	أتون الزجاج
collyrium of Isfahan	اثمد
basin	أجانة
black ink	احبار السود
colored inks	احبار الملونة
to open (the press)	ارخى
rice	أرز
to loosen	ازح
red lead	أسرنج
sponge	اسفنج
white lead	اسفيداج
plant carbonates	أشنان
awl	أشفة
finger (measure of length)	أصبع
spine (of a book)	أصل
white iqlimiya	أقليميا ابيض
vessel	أناء
copper vessel	أناء نحاس
opium	أفيون
acacia	أقاقيا
examination	امتحان
iron filings	برادة الحديد
to cool	برد
headband	برشمان
mace	بسباسة
to spread out	بسط
to dress (leather)	بشر
boxwood	بقس
sapanwood	بقم
oak trees, acorns	بلوط
violet	بنفسج
natron, borax	بورق
glair	بياض البيض
divider	بيكار
application of leather to the cover board	تبطين
bookbinding	تجليد
a stroke or curve executed on leather, etc.	تحنيش
bookbinder's press	تخت
to put into a press	تختيت
boarding (in bookbinding)	تركيب
impression	ترنجة
theriac	ترياق
to bind a book	تسفير
residue, precipitate	تفل
apprenticeship	تلعم
to be softened	تلين
the making supple of leather	تمحيط
dried pine cone	تمر الصنوبر اليابس
humidity	تنقيع
oven	تنور
Syrian mulberry	التوت الشامي

tutty	توتيا
floral or round ornament	توريق
to test, to experiment	جرب
rocket	جرجير
clay (?) vessel	جرّة
to write with liquid gold	جري
palm twig	جريد النخل
onyx	جزع
to dry	جفّ
leather, usually excellent leather	جلد
pomegranate flower	جلنار
to coagulate	جمد
wing, hinge	جناح
walnut	جوز
nutmeg	جوزة
lime	جير
side of a sheet of paper	جبهة
tannin ink	حبر
peacock-blue ink	حبر ازرق طاووس
myrobalan ink	حبر اهليلج
grand basil ink	حبر ريحاني
instant ink	حبر ساعتة
sunny ink	حبر شمس
peacock-colored ink	حبر طاووس
foreign ink	حبر غريب
rose-colored ink	حبر وردي
grain (a measure)	حبّة
to harden (as glue)	حثر
pumice stone	حجر القوصرى
polishing stone	حجرة ملسا
iron	حديد
iron tool used to reduce impressions	حديدة قاطعة
iron tool for decorative impression	الحديد للنقش
to shake, to stir	حرك
sieving cloth	حريرة
to sew the quires together	حزم
boxwood	حضض
to cut out	حفر
container (for dissolving glue)	حقة
to dissolve	حلّ
chick pea (a size)	حمص
sourness, acidity	حموضة
chalk	حوارة
mustard	خردل
to sew a skin	خرز
to undo the sewing (of a book)	خرم
St. John's bread	خرنوب
hardwood	خشب
grape vinegar	خلّ الكرم
wine vinegar	خلّ خمر
Chinese galanga	خلنخان
a perfume (includes saffron)	خلوق
sewing of a book	خياطة
thread on spine of a brochure	خيط

tepid	دافى
dāniq (a weight)	دانق
tanning	دباغ
tar soot	دخان النفط
soot of pine sap	دخان عقد الصنوبر
dirham (a unit of weight)	درهم
cover board of a book	دفّة
to pound	دقّ
to rub	دلك
color of the dragon's blood	دم الغزال
balm-tree oil	دهن البلسان
coconut oil	دهن النارجيل
inkwell	دواة
dissolved, melted	ذائب
to sprinkle	ذرّ
cubit (a unit of length)	ذراع
gold	ذهب
liquid gold	ذهب محلول
to melt	ذوب
handful	راحة
head (of a book)	رأس
filter	راووق
grape rob	رب عنب
marble	رخام
swelling	رخوة
to impress	رشم
Indian white tin	رصاص قلعي
to press (fruit)	رضّ
to be moist	رطب
raṭl (a unit of weight)	رطل
to thin (the leather)	رقق
parchment	رقوق
to set (the brochure in its cover)	ركب
to adjust the head and the side of the quires of a brochure	ركز
imprint left on a hide by a wedged tool, or the tool itself	ركن
ash	رماد
saliva	ريق
vitriol	زاج
green vitriol	زاج أخضر
green Cypriote vitriol	زاج أخضر قبرصي
yellow vitriol	زاج أصفر
Greek vitriol	زاج رومي
Iraqi vitriol	زاج عراقي
Persian vitriol (blue)	زاج فرسي
Cypriote vitriol	زاج قبرصي
emerald	زبرجل
dung	زبل
mercury	زيبق
red lead	زرقون
"red arsenic," realgar	زرنيخ أحمر
"yellow arsenic," orpiment	زرنيخ أصفر
saffron	زعفران
bitumen	زفت
a skin not depilated	زقّ
verdigris	زنجار
ginger	زنجبيل
cinnabar	زنجفر

flower of lead	زهرة الرصاص
olive oil	زيت
radish oil	زيت فجل
to pulverize	سحق
dung of asses	سرجين الدواب
to remove (from the hide's surface)	سرى
bookbinder	سفار
volume	سفر
to free a skin after tannage from foreign materials	سقع
to soak	سقى
white sugar	سكر ابيض
crystalline sugar	سكر طبرزد
knife for pen cutting	سكين البرى
knife for the nib	سكين القط
sorrel	سلق
fat of a cow	سمن البقر
sandarac	سندروس
small thong of leather	سير
sword	سيف
red lead	سيلقون
alum	شبّة
red alum	شبّ أحمر
dyers' alum	شبّ الصباغين
black alum of the dyers	شبّ الصباغين الاسود
"alum of the Carthamus," alkaline ash	شبّ العصفر
Yemenite alum	شبّ يمانى
yellow Yemenite alum	شبّ اليمانى الأصفر
fat	شحم

colocynth fat	شحم الحنظل
rope made of the palm tree	شريط
hair (used with clay for luting)	شعر
barley	شعير
paring knife	شفرة
split (of the pen)	شقّ
to weigh	شقل
anemone	شقيق
soap	صابون
adherence	صاق
to preserve	صان
to decant, to pour	صبّ
shallow basin	صحن
willow	صفصاف
to purify	صفى
to polish	صقل
a hard stone on which pulverization is carried out	صلاية
gum of scammony	صمغ السقمونيا
gum of acacia	صمغ القرض
gum arabic	صمغ عربى
sandalwood	صندل
wool	صوف
compass	ضابط
to hold with force	ضبط
thickened (pasteboards)	ضخامة
to beat (in order to mix)	ضرب
pressed	ضغط
to put together the edges of a hide	ضمّ
to decoct, to cook	طبخ

edge (of a cover board)	طرف
flat dish	طست
to coat with glue or milk	طلا
mica, talc	طلق
to fold	طوى
glazed pot	طيجن مطلى
copper pot	طيجن نحاس
to lute (a vessel)	طين
ivory	عاج
to knead	عجن
tamarisk gall	عدبة
tendons (for glue preparation)	عراقب
celandine	عروق الصباغين
honey	عسل
extract of myrtle	عصارة الآس
to press	عصر
Carthamus	عصفر
to fold	عطف
gallnut of the terebinth	عفص البطم
Greek gallnut	عفص رومى
to deteriorate	عفن
catchword collation	عقب
to thicken, to become more viscous	عقد
to make a distinctive sign	علم
to shape (leather)	عمل
ambergris	عنبر
onion	عنصل
Indian aloeswood	عود هندى
dust	غبار

glue	غراء
snail glue	غراء الحلزون
fish glue	غراء الحوت
fish glue	غراء سمك
Syrian glue (a fish glue)	غراء شامى
a sieve	غربال
to wash (minerals, chemicals)	غسل
to boil	غلى
charcoal	فحم
furnace	فرن
turquoise	فيروزج
to crumple a hide between the fingers	فرى
pistachio	فستق
to cut out	فصل
silver	فضّة
glass vessel with a wide mouth	فقّاعة زجاج
cheek of the press	فلق
mouth (of a vessel)	فم
sharp	قاطع
mold, mold size	قالب
quarto	رباعنى
octavo	ثمانى
duodecimo	زلبجة
beaker	قدح
glass beaker	قدح زجاج
pot (used for glue)	قدر
lead pot	قدر رصاص
carat (a unit of weight)	قراط

a whitish color	قرش
to put into the press	قرص
to trace	قرض
paper	قرطاس
to square, bevel square	قرطبون
distillatory, alembic and ambix	قرعة وانبيق
clove	قرنفل
flask of glass	قرورة زجاج
tin	قزدير
costus	قسط
rinds of pomegranate	قشور الرمّان
to trim the edges of a book	قصص
nib of a pen	قطّة
to drip	قطر
bottom of a bowl	قعر
spine of a book	قفا
blue vitriol	قلقنت
pen	قلم
pen of the third	قلم الثلث
pen of the two thirds	قلم الثلثين
quill pen	قلم الرياش
hair pen	القلم الشعر
pen for capital writing	قلم الطومار
pen of the half	قلم النصف
clay vessel (for the furnace)	قلة
soda	قلّى
flask (which may be heated)	قمقم
glass flask, phial	قنينية
thin walled vessels (buried in dung)	قوارير دقاق

viscous	قوام
to cover with gold using a pen	كاسف
paper	كاغد
sulfur	كبريت
tragacanth gum	كثيرا
collyrium	كحل
leek	كرّاث
stitched paper book	كراس
coriander	كزبرة
to cover with leather	كسا
fennel, seseli	كلخ
folding	كماش
frankincense	كندر
laudanum	لادن
lazward, lapis lazuli	لازورد
milk	لبن
woman's milk	لبن النساء
cow's milk	لبن بقر
yoghurt	لبن حليب
milk of the wild ass	لبن الحمر الوحشية
camel's milk	لبن ناقة
tongue of a book	لسان
to glue	لصق
lac	لك
red lac	لك أحمر
wooden plate on which one cuts out the cover boards	لوح
almond	لوز
līq	ليق
pad of wool to receive the ink	ليقة
to soften	لين

water or juice of rush	ماء الأسل
juice of the gum	ماء الصمغ
juice of bran	ماء النخالة
rosewater	ماء الورد
sweet water (i.e. unsalted water)	ماء عذب
juice of camphor	ماء كافور
lemon juice	ماء ليمون
tool	ماعون
knife for cutting out	مبزق
twisted	مبروم
ox bladder	مثانة ثور
mithqāl (a unit of weight)	مثقال
brazier	مجمر
shell fastened in a handle used to polish applied gold	محارة
to make leather supple	محط
Prunus mahaleb L.	محلب
dissolved (of alum, gum arabic, etc.)	محلول
to rub, to polish, also an iron tool in the shape of a gendarme's hat used to trace the marking on leather	مخط
soot ink	مداد
Chinese ink	مداد صيني
Iraqi ink	مداد عراقي
Persian ink	مداد فرسي
Kufic ink	مداد كوفي
compound ink	مداد مركب
Nafuran ink	مداد نفوراني
India ink	مداد هندي
to stretch skin by keeping it between boards	مدّ
myrrh	مرّ

goat buck's gall	مرارة تيس
part of the cover destined to protect the foredge	مرجع
decorated, impressed	مرشوم
crushed	مرضوض
marcasite	مرقشيتا
paste	مرهم
straightedge	مسطرة
equalized	مستوى
to test a skin by passing the hand over it	مسح
to seize	مسك
musk	مسك
apricot	مشمش
distillate, sublimate	مصاعد
glass burnishers	مصاقل الزجاج
mastic gum of the lentisk	مصطكى
hammer	مطرقة
manipulation, treatment	معالجة
press	معصرة
spinner's press	معصرة المغازل
glued	معقود
iron spoon	مغرفة حديد
a tool to cut out a hide or paper	مغرط
fleshed (of a hide)	مقشر
scissors	مقص
base on which the nib is cut	مقط
false bdellium	مقل
blue bdellium	مقل أزرق
table salt	ملح الطعام

Andāranī salt	ملح اندرانى
leather sieves	مناخل الجلود
wooden mallet	منجم
miller's sieve	منخل قاروط
iron mortar	مهراس حديد
storax	ميعة
to arrange pagination in order	ناسب
a mixture of equal parts of lazward and *baurūq*	نجوبى
copper	نحاس
to scrape	نحت
to sieve, to filter	نخل
hair sieve	نخل شعر
to weave	نسج
starch adhesive	نشاستج
to spread a skin	نشر
to absorb (as of water or glue which penetrates paper or a hide)	نشف
white *nafṭ* (*or nifṭ*), a petroleum product	نفط ابيض
a tool for impression, to impresss, to decorate	نقش
to macerate (a medicine) in water	نقع
small tool for stamping leather	نوارة
sal ammoniac	نوشادر
indigo	نيل
Indian indigo	النيل الهندى
water lily	نيلوفر
cardamom	هال
mortar (for pulverization)	هاون
to decant, to pour out	هرق
to press out	هشم
side of a sheet of paper	وجة
tamarisk leaf	ورق الأثل
gold leaf	ورق الذهب
sheet of paper or gold leaf	ورقة
middle leaf (of a quire or book)	الورقة الوسطى
gum ammoniac	وسق
to embellish (a book by impression and stamping)	وشح
ounce (a unit of weight)	وقية
red ruby	ياقوت

GREEK INDEX

INDEX OF FOREIGN WORDS OTHER THAN GREEK

The particular languages are denoted in the text and notes.

GENERAL INDEX

Gilding, 51ff.; gold substitute, 39; mediaeval, 9
Ginger, testing of, 49
Glair, 7; for dry ink, 19; in ink, 7; to temper blues, 8
Glass, for black ink, 19; ground for ink, 35
Glass burnisher, for paper, 40
Glass dyes, 31
Glass pot, 18
Glass vessel, 39, 46
Glassmaking furnace, 17, 35
Glazed pan, 11, 35, 40
Glazed paper sheet, 35
Glazed pot, 11
Glazing of paper, 10, 40
Glue, 9; excessive, 53; from fish, 37; from hide scraps, 9; from snails, 37; in a brownish color, 31; in a substitute silver ink, 33; in Sumer, 45; its solution, 37; manufacture, of, 37; shopkeeper's, 54–55; solution from leather scraps, 45; Syrian, 54–55; with unslaked lime for ink, 9; work with, for gold adherence, 37
Gluepot, 11; description of, 37
Gluing, knife for, 38; of gold and silver, 9; of the quires, 43
Goffer, 53
Goffering tool, 53
Gold, adherence of, 9, 54; coloring for, 31; Ibriz, 37; pounding of, 54; pulverization of, 32; pulverized for ink, 9; purification of, 24; rubbing, 55; solution, 54; washing, 54
Gold color, 8
Gold colored ink, 25
Gold filings, 32; in a *līq*, 29
Gold ink, 9
Gold *iqlimīyā*, 36
Gold leaf, 32; Ibriz, 37; rubbing, 54
Gold *līq*, 54
Gold polishing agents, 38
Gold scrapings, for gallnut ink, 18
Gold size, 9
Gold substitute in gilding, 39
Gold writing, sulphur for, 33
Golden ink, 19, 27; for parchment, 22; preparation of, 25
Golden *līq*, preparation of, 28
Goldwriting, preparation for, 32
Gotha manuscripts, 6, 7
Gouge, for decorating, 42
Gout, colocynth for, 34; henbane for, 50; willow for, 38
Grand basil ink, preparation of, 23
Gratzl, E., 5
Greater celandine, 26
Greater galingale, 31
Greek bean, 28
Greek gallnut, as a febrifuge, 16; for Kufic ink, 16; in primary ink, 36
Greek lovage, 34
Greek mustard, 34
Greek verdigris, 47; testing of, 46
Greek vitriol, 32; for gallnut ink, 18; for peacock colored ink, 22
Green coriander juice, 31
Green gallnut, for dry ink, 20; for sunny ink, 19; for ink for the common people, 19; in Syrian mulberry ink, 20
Green gum, for dry ink, 19
Green ink, 21, 32; preparation of, 23, 25
Green *līq*, 27; preparation of, 29
Green pigment, 8
Green soot, from sulphur, 40
Green vitriol, 16; for dry ink, 20; for ink for the common people, 19; for instant ink, 19; in myrobalan ink, 19; known to Babylonians, 19
Grohmann, A., 6
Gum, from acacia, 17; green for dry ink, 19; in gold size, 9; in ink, 8; purpose of, 46; tempering, 8; to retain viscosity, 8
Gum ammoniac, for a golden ink, 23–24; medicinal uses of, 24
Gum arabic, 7, 15; for dry ink, 19; for instant ink, 19; for red ink, 22; in a medicine, 13; in gallnut ink, 18; in Iraqi ink, 17; in preservation of *līq*, 18; in silver metal ink, 33
Gum arabic solution, preparation of, 28
Gum of *el-kelkh*, 24
Gum of scammony, 15
Gum of the *ṭalḥ*, 15
Gum resin, 30; from scammony, 15
Gum solution, for India ink, 16
Gum tragacanth, as a tempera, 8; in a red *līq*, 29; in a white *līq*, 29; in ink, 8; vehicle for metal, 9
Gums, pomegranate flower for, 26

Hadad, H., 6
Hadramut aloes, 50
Hainan, 31
Hair, in clay of the art, 17
Hair brushes, for writing, 37
Hair dye, 16, 24
Hair sieve, 34
Hairy saffron, 32
Half-Manṣūrī size, 41
Hamadan, 16
al-Hamdānī, 6
Hamilton, G. H., 7, 8
Hammer-Purgstall, J., 6
Hammurabi's Code, 37
Handful, a measure, 37
Hara nut, 44
ibn Ḥassan, 24
Head, mustard for, 28
Headache, anemones for, 17; boxwood for, 42; henbane for, 50; scammony for, 15
Headband, work with, 51ff.
Hearing difficulty, sulphur used for, 33
Heart, orange used for, 30; sandalwood for, 38
Heart palpitation, clove for, 48; use of lesser galingale for, 31
Heat, for sympathetic ink, 9
Heating apparatus, 17
Hebrew, 16
Helix, 41
Hemorrhage, burned sponge for, 38; internal, use of sandarac for, 17; talc used for, 24; uterine, sumac for, 24
Hemorrhoids, leek used for, 27; mustard for, 28; myrtle for, 21; onion for, 36
Henbane seed, 50; testing of, 50
Hepatitis, *Indigofera* for, 23
Herpes, use of sulphur for, 33
Herpetic eruptions, sponge for, 38
Herringham, C. J., 10
Hibiscus cannabinus L., 39
Hiller, J. E., 15
Hinges, 43; of book, 52
Hippocrates, use of sulphur by, 18
Holmyard, E. J., 8
Honey, for compounded ink, 46; for gold *līq*, 29; for gold size, 9; in golden ink, 27; in medicine, 24; in bread, 24; to make alcoholic drinks, 24; with vinegar, 24
"Horny matter," 49
al-Ḥusain ibn Ṣaffār, 6
Hydragogue colocynth, 34
Hyoscyamus albus L., 50
Hyoscyamus aureus L., 50
Hyoscyamus niger L. 50
Hypnotica, saffron in, 22
Hysteric depression, saffron for, 22

Ibrāhīm, 6
Iliad, 51
Illumination, inks for, 7; materials for, 8
Immediate ink, 7
Impetigo, use of walnut for, 21
Incense, 9, 45
Incontinence, of urine, use of aloeswood for, 38
India ink, 7; preparation of, 15, 16
Indian aloeswood, for the pen, 38
Indian buckthorn, 48
Indian indigo, a basic color, 29; for a *lāzward līq*, 27; for a red *līq*, 26
Indian musk, testing of, 47
Indian myrobalan, 44
Indian salt, 32
Indian soot, manufacture of, 7
Indian spikenard, in dyeing of leather, 44
Indian *tanbitī*, testing of, 47
Indigo, for a blue color, 32; for a pistachio color, 31; for a violet color, 32; for green ink, 23; in a green *līq*; in tinting, 9; Indian, for a red *līq*, 26; Iraqi, for a yellow *līq*, 27
Indigo juice, for a blue *līq*, 27
Indigo plant, 23
Indigofera tinctoria L., 23
Inflammations, henbane for, 50; sumac used to prevent, 24
Ink, black, 7; black and white, 24; blue-black, 7; cheap, 7; chemical, 13; chemically active, 8; Chinese, 7; dry, 7; Egyptian, 46; essence, 18; for common people, 19; for monks, 25; for paper sheets, 34; for papyrus, 7; for religious books, 20, 31; for travelers, 7; from anemones, 23; green, 23; gold, 9; India, 7; instant, 18, 22; Iraqi, 7, 17; Kufic, 7; lead, 9; metallic appearance of, 24; myrobalan, preparation of, 19; Nafuran 7, 17; parchment, 22; peacock color, preparation of, 22; Persian, 7, 17; primary, 9; purple, 23; reaction with parchment, 13; red, yellow, and green; 21; rose colored, 23; secret, 9; silver, 9; soot, preparation of, 15; tannin, 7; tin, 9; white, preparation of, 25; with glass particles, 35
Ink powder, preparation of, 20
Inking, straightedge for, 42